T0064186

Zoltan J Kiss

Matter =
= the Matrix of Information

2017

Order this book online at www.trafford.com
or email orders@trafford.com

Most Trafford titles are also available at major online book retailers.

Print information available on the last page.

ISBN: 978-1-4907-8417-5 (sc)
ISBN: 978-1-4907-8416-8 (e)

Trafford rev. 08/23/2017

www.trafford.com

North America & international
toll-free: 1 888 232 4444 (USA & Canada)
fax: 812 355 4082

for all those who trust in renewal

and want to see the proofs…

EXECUTIVE SUMMARY

"There is no *matter* as such" told *Max Planck* and named *God,* as the personalization of the intelligent mind, who holds the world together – as conclusion of his speech on "The Nature of Matter" conference at Florence in 1944. While his explanation was pretty global, he has given also the point, as he added: *"all matter originates and exists only by virtues of a force which brings the particle of an atom to vibration and holds this most minute solar system of the atom together.*"

This is in fact all about the process based elementary approach.
This is all about processes in the elementary world rather than particles. The tools of the intelligent mind are no others than events in time, intensities, temperature, quantum communication, quantum impulses, quantum impacts, space, space-times, gravitation, quantum speed and similar categories. The *minutest solar system* is based on balanced quantum communication – permanent change between the two *inflexions* of the elementary cycles.

The anti-proton/proton *inflexion* is the origin of the proton process, the sphere symmetrical expanding acceleration, *the capacity of the intensities*, accumulated during the anti-proton collapse. The proton process is quantum impact of certain *constant* number of quantum impulses for certain periods, increasing in time, with decreasing frequency. The absolute value of the intensity capacity available is without change, but with the frequency of the impact is getting less and less in time.
The electron process is *work intensity at constant time system*, but with the number of the impacted quantum decreasing in time. This work intensity drives the neutron process into collapse. The number of the impacted quantum is less and less, until the electron process has its intensity capacity to drive the collapse of the neutron process.

The expansion is a process, happening on its own, with the use of its own energy. The collapse needs external drive. The drive of the neutron collapse is the electron process, still part of the expansion. The proton process is quantum impact; equal number of impacted quantum for periods extending in time; the transformation of energy. Transformation of the quantum impact (energy) from high intensity to low intensity.

The electron process means external work at constant time system; quantum impact with decreasing number of impacted quantum.

The neutron process needs external energy, the work of the electron process, for driving the collapse, and this way generating the capacity of the neutron collapse with increasing intensity.

The elementary process is the connection between the proton and the neutron processes, between the expansion and the collapse. The intensity, developing during the proton process is transported to the neutron process for making the elementary balance to happen. The intensity of the energy during the proton process is decreasing. The intensity of the neutron process, during the collapse, driven by the electron process, is increasing.

The neutron/anti-neutron *inflexion* is the change of the direction of the elementary process. The anti-neutron process is expansion, repeating the accumulated by the collapse intensity of the neutron process. The anti-neutron process re-generates the energy/intensity of the neutron collapse. The anti-electron process controls the elementary cycle and drives the elementary evolution from the *plasma* to the *Hydrogen* process – the source of *gravitation*.

The anti-neutron process is similar to the proton process, as all its energy intensity is developing at the moment of the inflexion and this high intensity impact is decreasing during the expansion. The number of the impacted quantum at the inflexion remains the same, but the duration of the impact is increasing: the intensity of the generating energy is getting less and less – the frequency of the impact is decreasing.
The anti-electron process is the drive of the anti-proton process, the collapse on the anti-side, bringing the elementary process back to the start of the next elementary cycle.

The global lesson of the operation of an elementary cycle is that there are two energy/intensity exchange processes within the cycle: one on the direct side and the other one on the anti-side. The direct side is responsible for the elementary communication, the anti-side for the control of the elementary process.

The intensities of the anti-electron processes of all elementary processes are equal. The difference between the anti-electron processes is in the duration of their driving impacts. Elementary processes with higher intensities at the inflexions need/have the anti-electron process drive for longer duration. The equal intensities and the difference in the durations result in the surplus of quantum impacts on the anti-side of the process. The surplus of this quantum impact is the origin of the quantum impact of *gravitation*. The higher the periodic number of the elementary process is, the higher is the absolute value of the contribution of the elementary process to the quantum impact of *gravitation*.

The progress of the elementary evolution means the periodicity of the elementary processes with different and decreasing quantum speed and intensity values and with the generation of the quantum impulse.
Electron process quantum drives cannot run out to zero intensity, as in this case there would be no intensity/energy for the collapse to start from. The “nothing” (zero intensity, fully expired) cannot be the starting point of the next elementary step. Therefore, all elementary processes have a remaining, residual quantum entropy value within the electron process. The power, the quantum impact capacity of this residual entropy value is of infinite small, not capable for any drive, but still not zero. This remaining quantum impact entropy value is the quantum impulse (quantum), the building stone of the *space*.

The decrease in the quantum drive capacity of elementary processes has its influence on the quantum speed values of their operation. This is the reason the quantum speed values of elementary processes vary. The speed value of the quantum communication is the characteristic of the space-time of the elementary process. The higher the value of the quantum speed is, the higher is the starting intensity of the proton process; the higher is the number of the impacted quantum; the wider is the space-time of the elementary process. The quantum speed of the elementary processes becomes cycle by cycle less and less. Here, in our *Earth's* space-time the value of the quantum speed corresponds to the speed of light.

The generating in the elementary processes quantum impulses (quantum) compose the quantum space. In other words: The quantum space (or just space) has been built up from the generating during the elementary processes residual quantum entropy impact values – quantum impulses (quantum). This obviously means: There is no space without elementary processes and there are no elementary processes without quantum space.

The *space* and the *space-time* are different categories. Space is composed from quantum impulses. Space-time is the characteristic of the process with certain acting quantum speed value and time count. Space-time is the category of the elementary communication. If the space is the garden than the space-times are the trees within the garden.

Events happen in space and time. Definition of *time* cannot be given without event and *space* is composed from quantum impulses, the entropy remains of elementary processes (events). Each elementary process has its own space-time with the specific speed and intensity value of the quantum communication. High intensity quantum drive means wide space(-time) and quick time flow, with a lot of events. Low intensity quantum drives operate within smaller space(-times) and have increased time count, with low number of events. Events happen in the variety of the space-times in parallel. One and the same event may appear in different space-times for different time counts. In high intensity space-times the time is ticking slower. The time count difference between the space-times of the elementary processes of our Periodic Table is not significant. The value of the relativistic coefficient, from the *Helium* process to the trans-*Plutonium* processes corresponds to 0.98 – 1.65. *Plasma* has infinite large space-time, with infinite high intensity of the process; while the *Hydrogen* process has infinite small space-time and infinite long process time duration. We on the *Earth* surface live somewhere between the two.

All electron and anti-electron processes operate at the same time system. This is the basis of the elementary communication. The common time system provides however the common platform not just for the communication, but also for the conflicts as well. And the *blue shift* conflicts of electron processes create different aggregate elementary stages.

Conflicts with high intensity results in a kind of plasma status. The higher the number of the participating in the conflict electron process *blue shift* impacts is, the higher is the conflict itself. The measurement of the conflict is the *temperature*. Higher conflict results in higher intensity impact, in higher temperature.

High intensity conflict can be cooled down, in which case, the intensity is decreasing, while at the same time the number and the origin of the impacts may remain the same. High intensity conflicts with massive cooling effect may result in quantum communication of decreased intensity status: in liquid, gaseous or solid aggregate statuses. The origin and the number of the quantum impacts remain the same just the symptoms of the conflict become modified – result of the cooling impact. Without taking away quantum impacts, without cooling, the conflict is of high intensity; in the case of cooling the impact is different, resulting in lower intensity status.

The cooling impact of the *Earth* plasma is natural. The high quantum speed within the centre and its decreasing value towards the surface well represent the difference in space-times and time counts. The space-time of the *Earth's* plasma is infinite large, with quantum speed of infinite high value. Our space-time on the *Earth* surface belongs to our quantum speed value on the *Earth* surface, the speed of light.
These are important circumstances, as the time count on the *Earth* surface and the time count of the plasma process are pretty different. The event, happening in our time count (in our space-time) might be infinite long, while for the plasma it might be just a "sneezing". The conflict in the centre of the *Earth* results in plasma. The same conflict, just cooled down on the surface corresponds to the solid aggregate status of the crust.

The global collapse of the neutron/electron processes of all accumulated *Hydrogen* processes results in plasma, as this collapse generates conflict of infinite high intensity. Infinite high quantum speed results in infinite large space-time, in infinite high intensity and in infinite short acting time. With the cooldown of the conflict of the plasma the number of the participating in the conflict elementary *blue shift* (quantum) impacts do not change. The intensity of the quantum impacts of the conflict is the one, which becomes modified. And we find lava, gaseous and solid aggregate statuses instead of plasma with infinite high intensity.

This is all about quantum impacts.
The intensity of the anti-proton/proton inflexion characterises the quantum speed and the number of the impacted quantum as well. The intensity level of the proton process in our time system is well represented by its estimated lifetime of 10^{29} years. The infinite high intensity impact of the plasma status has its duration in our *Earth's* space-time, of infinite length. The same impact, which appears as being fluent lava at certain circumstances, is solid mineral in our life. The difference is in the time counts, in the intensity status. The process is the same, the appearances are clearly different.
With the generation of conflicts, solid subjects can be melted, liquids turned into gaseous state, gases exploded by the increase of the pressure, the conflict.

Yes. *There is no matter as such existing*!
All what we find around us are processes, with aggregate status, function of the intensity of the internal elementary quantum communication of the components!

Matter in fact is no other than quantum communication, result of the *blue shift* quantum impact of the electron processes – in very simple terms: *the matrix of information.*

This finding correlates well with the official standard model. The "*force which brings the particle of an atom to vibration*" is uniform. The operation of the strong and the weak interrelations, the *gravitation* and the electromagnetic impacts are of one and the same rules and principles. There is no need to separate events from events, processes from processes, classify different relations only for making explanations on their own. The world is not divided, is not for separation.
The world of the "*intelligent mind*" as of *Planck's* scientific legacy is homogenous.

The process based approach globalise the view and grants the necessary harmony to the measured experimental data and the global theory, the understanding of our world: the quark processes represent the different intensity stages of the elementary expansion and the collapse; the *W* and the *Z* bosons represent the intensities of the inflexions; the anti-electron processes are the origins of the electromagnetic impacts and the quantum impacts of gravitation; *gravitation* is part of the elementary quantum communication.

Quantum impacts are signals propagating within the quantum space in any direction. The quantum impulses (quantum) of the space transfer all quantum impacts without any modification. All quantum impacts are electron process *blue shift* impacts. The intensity and the conflict of these quantum impacts is the one, establishing the aggregate status of the elementary processes.
Quantum impacts of infinite high intensity and conflict result in the plasma status.
Quantum impacts of "moderate" intensities and conflicts result in gaseous, liquid and solid aggregate statuses. The intensity and the conflict of the quantum impacts (of the electron and the anti-electron processes) establish the quantum speed value of the communication of the elementary processes. As the intensities of the quantum impacts of the communication and the intensities of the developing conflicts are the ones establishing the form of the appearance of the event – *matter is the matrix of the information* indeed, with *infinite large variety* of options, available.

The characteristics of the inflexions, the balance of the processes and anti-processes, the *blue shift* impact of gravitation, the process based interpretation of the standard model are given; the sphere symmetrical expanding acceleration of the *Earth* is proven; the experimental proof of the conflict of the *Hydrogen* process in acceleration and the quantum impact of *gravitation* is presented; the *matter* is the product of the quantum impacts of elementary processes, the result of the elementary communication, *is a kind of complex information about the event*!

The understanding of the world around us raises at least one global question: is the world with an end or without?
While the particle based approach might be the symbol of the world with an end, answering this question on process basis, it gives the opposite result for sure: Infinity has been embodied by processes. The world is without end; the definition neither of the biggest

nor of the smallest can be given. Zero is not about the end of the road, rather the point of the *inflexion*: the milestone of the change of the direction. There will always be values to be found closer and closer to zero, or to a selected number; bigger than the biggest and smaller than the smallest.

Our world has been built up on infinite number of conflicting couples for finding the natural balance between them all. There is no small without big, no empty without full; the speeding up loses its meaning if the slowdown is missing. There is no short without long, as there is no forward without backward…The list is without end.

Communication is the basic principle of our existence. Nature is not about self-separation, excluding others from our life, rather offering our strengths to others and asking for support where we need it. "I could not feel well", the *Oxygen* process could reveal "if I could not communicate with others, if I could not help those who need me. I would not survive alone as *Oxygen*, fighting others with all the energy of mine, instead of fully communicating. My life would be over senselessly."

Differences are the symbols of the freedom. Conflicts are the drives of the progress. The balance is the solution. The permanent progress means the elementary evolution.

Events happen in time. There is no event without time and there is no time without event. The only question is: what the intensity of the event is? If the intensity of an event is infinite low, it might be taken as something without any change, while in fact the only reason of this assumption is the infinite long duration of the event. The variety of the intensities is infinite large and this way the number of the existing time system options is also infinite large. The same event may happen in different space-times and therefore for different durations.

We are parts of the big whole and the intensity of our system depends on the speed value of the quantum communication of our space-time. This quantum speed in our *Earth's* space-time is the speed of light. Elementary processes mean different space-times. Different space-times mean different speed values of quantum communication.

The official definition of the speed of light, as the speed of photons in empty space shall be revised. This is not just about the double, particle and wave based nature of the supposed photon in motion; but "empty space", as such has no meaning and the definition with vacuum, or with the almost vacuum does not sound better as well. In addition the definition of space and gravitation are also missing as official categories.

There is no need to associate the particle based official view with the world with an end concept for any costs, but in the case of processes there is no problem with the definition of time: events happen – the time is ticking. The time flow and the end of the world as such however are in clear contradiction!

Science always was the best example of the freedom of minds. The thoughts about the world with an end or without are to be discussed. The same way as the time, the space, the gravitation, the temperature and the matter as well. This is simply the demand of the day.

If the change is permanent, the world is without an end. There is no way to define any particle within, which would symbolise the status of rest in its any formats, whatever small it is considered to be, not just because there always will be a smaller, but for the reason, there is no way to expect the process stops: No event would mean no time.

The sliced model in physics generates sliced rules in their mathematics as well. But there is no way to start to specify each slice on its own, just because the global approach is missing. Is there either a uniform way to describe our world is it big or small, sweet or bitter… or there is no other way. The process based approach provides the global view!

Matter, as matrix of quantum impacts (information) is far not just an elementary category in theory only.
Air, water, hydrocarbons, minerals, all our natural and man-made products are results of the quantum communication of the elementary processes. The form of their appearance is function of the intensities, the values of the quantum speed, the internal and external conflicts, the elementary balance – function of the overall conditions of the quantum communication. The thermal status of the pyramid, for example suffers immediate change, once its internal energy status becomes impacted; or concrete structures become water tight and close-non-breakable if the internal balance is good enough.
Quantum drives of increased intensities keep elements (elementary processes) at permanent temperature difference from each other and from the environment.
The eight elementary processes (*H, He, C, N, O, Si, S, Ca*) with electron process surplus are the initiators of the quantum communication; the others with increased intensities welcome the communication, share their energy potential, producing altogether the elementary/mineral wealth of the world we live in.
Matter – at the end of the day – is the materialisation of quantum impacts!

Gravitation, the product of the elementary evolution is our free energy source!

Our energy generation technologies are today in the double trap of efficiency and environmental impact. Efficient forms are all with significant environmental impact, while all those, without, or with less environmental impact are of low or limited efficiency.

The conflict of the *Hydrogen* process in acceleration and the quantum impact of gravitation generate heat. The principle of the energy generation is simple: The "instrument" is the *Hydrogen* process, speeded up, the source is *gravitation.* The measured results of the small scale experiments, given in the book prove the potential of this technology.

\- * -

The sections in the book try to give full information on the discussed subjects in each case. But this is the sixth book with the general content of relativity, elementary processes, quantum impact, quantum impulse, gravitation, space, energy balance, quantum speed, space-times, etc. Therefore it might happen that while references are given the best possible way, the explanations still need additional arguments and mathematical proofs, which are given in the previous books.

These books are:

1. The Energy Balance of Relativity (2007)
2. Quantum Energy and Mass Balance (2009)
3. Quantum Engine (2011)
4. Gravitation: our quantum treasure (2013)
5. The quantum impulse and the space-time matrix (2015)

As for specific interest,

- the comments/notes

  to *Einstein's* works on moving bodies (1905), on the propagation of light (1911) and on the special and the foundation of the general theory of relativity (1916)

  are also given in this book, as Attachment. [They were part of the book for "The Energy Balance of Relativity." 2007]

  The note states:

  Einstein's assessment is not full; important relativistic aspects have been missing!

  In the case the review of the subjects is made by the necessary completeness, the conclusions are different.

- Section 4.6, part of Chapter 4 for "Elementary communication in practice" gives the proof of the sphere symmetrical expanding acceleration of the *Earth*, the reason of *gravitation.*

Table of Content

1	***Elementary processes* and the *quantum impulse***	1
1.1	The meaning of the quantum speed	4
1.1.1	What the *quantum impact of gravitation* (*qig*) is about in spaces above certain structures, within closed premises and in the vacuum chamber?	8
1.2	Additional arguments, explaining the variety of the quantum speed values in elementary processes	8
1.3	The elementary processes are about *change*	12
1.4	The *quantum impulse* (quantum) and the elementary balance	13
1.5	The generation of the *quantum impulse*	16
1.6	The common space-time of the proton and neutron processes	18
1.7	The relation of the *quantum impulse* and the size of *space*	22
2	**Inflexion**	23
2.0.1	*Comment to the inflexions*	26
2.1	The *proton processes* are similar, with the appearance of this similarity within different space-times with different quantum speed and electron process intensity values	27
2.2	The quantum impact of the motion	29
2.3	Parallel events with different time counts	32
2.3.1	In addition to the subject	35
3	**Elementary communication**	37
3.1	The relation of the *quantum impulse* and the intensity of the elementary process	38
3.2	The sequence of elementary cycles	42
3.2.1	Proton process dominance	43
3.2.2	Neutron process dominance	44
3.2.3	Each elementary process has its own space-time	44
3.3	The meaning of the *parallel* and the *internal* elementary cycles	45
3.4	Elementary communication	52
3.4.1	*Water*	53
3.4.2	*Steam*	55
3.4.3	*Hydrocarbons*	56
3.5	The impact of parallel cycles	56
3.6	The net of elementary processes	59

4	**Elementary communication in practice The benefit of the quantum impact**	62
4.1	The example of the *communication*: concrete structures	64
4.2	Temperature means the *conflict* of the electron process	69
4.2.1	*Does the intensity increase modify the time flow?*	70
4.3	The quantum communication of the *pyramid*	71
4.3.1	Pyramids are in quantum communication with each other	76
4.4	The quantum impact of *gravitation* on convex/concave surfaces	77
4.5	Convex/concave structure with *Hydrogen* process inside	80
4.6	*Earth's* expansion	81
4.7	Specific conditions of *Earth's* expansion	85
4.8	Ice in its many formats	86
5	**The Standard Model with process based argumentation**	88
5.1	The space-time of the proton and the neutron processes	91
5.2.	*Intensities* instead of mass	92
5.3	The *proton* and the *neutron* processes – the conformity of the measured experimental data with the process based approach	93
5.4	*Bosons*, the quantum impacts (and about the *weak interactions)*	97
5.5	*Electromagnetic* interrelation	99
5.6	*Gravitational* interrelation	102
5.7	*Photons* versus *quantum impulse*	106
6	***Isotopes* and their rehabilitation**	109
6.1	*Beta* radiation means damage in the electron process	110
6.2	*Alpha* and *gamma* radiation	112
6.3	*X-ray* and *neutron* radiation	113
6.4	Rehabilitation of isotopes	113
7	**The *space* and the *space-time***	116
7.1	*The space*	118
7.2	The distance, quantum impacts of different space-times make	122
7.3	Communication of space-times	125
7.4	Computer communication	127
7.5	Source of energy generation	129
7.6	The space is one and the same and space-times contain each other	130
8	**The *matter***	133
8.1	Once again about the *matter*	135
8.2	*Gravitation,* the formulation of the *matter*	137
8.3	Soil and ash	140
8.4	*Matter* is the matrix of information	141

9 The free energy is with us 143
9.1 Introduction, short explanatory note about the principles 143
9.2 Examples of the small scale *Hydrogen* process acceleration experiment 148
9.2.1 Additional principal notes to the experiment 149
9.2.2 Results with the two-armed balance 151
9.3 The conclusion on the measurements 155

A **Attachment**
Review of Time and Space Coordinate Relations of the Special Theory and the Foundation of the General Theory of Relativity 159
10.1 Review of the theory on moving rigid bodies and moving clocks 159
10.2 Concerns about *Einstein's* formula 166
10.3 Review of the statement on *Gravitation* of Energy 167
10.4 Review of the principles of the foundation of the General Theory 172

Ch.1

1
Elementary processes and the *quantum impulse*

The <u>*proton process*</u> is sphere symmetrical expanding acceleration, from $v = 0$ up to speed *i*:

$$\frac{dmc^2}{dt_o}\left(1-\sqrt{1-\frac{v^2}{c^2}}\right)=\frac{dmc^2}{dt_o}-\frac{dmc^2}{dt_i};$$

where $i = \lim a \cdot \Delta t = c$; and dt_o represents the time count of rest within the space-time of the elementary process; dt_i corresponds to the speed value of *i*. 1A1

dt_o is relating to a *certain number of quantum impulses, quantum*, impacted closest in time to the anti-proton/proton process *inflexion*. While the *inflexion* itself is of *zero* time count, the start of the proton process shall correspond to this closest to the inflexion "at rest" status ($\lim \Delta t_o = 0$), otherwise formula 1A1 above has no meaning.

$$e_x = \frac{dmc_x^2}{dt_{ox}} \approx \frac{dn_x}{dt_{ox}} \approx f_{ox};$$

The generating impact at the *inflexion* corresponds to the full energy/intensity potential of the elementary process. 1A2

The energy/intensity potential of the inflexion means the impact of the full number of the quantum of the expansion. This is the reason this full impact corresponds to the *frequency* value in 1A2 above. This is the intensity source/demand of the whole elementary process, generating in its full value at the *inflexion* – the entire energy/intensity of the expansion! The total number of the impacted quantum at the rest status is proportional to the quantum speed value of the inflexion.

The higher the number of the impacted quantum is, the higher is the intensity and the quantum speed of the impact: $e_x \approx c_x^2 \approx \dfrac{n}{\Delta t_o};$ 1A3

There are *two* options above here, related to 1A2: Ref. 1A2

either, (α)	or, (β)
$e_x = \dfrac{dn_x}{dt_o = const};$ if $n_{x1} > n_{x2} \rightarrow f_{o1} > f_{o2}$	$dt_{ox} = \dfrac{dn = const}{e_x};$ if $e_{x1} > e_{x2} \rightarrow f_{o1} > f_{o2}$

1A4

<u>The higher the energy/intensity of the *inflexion* is</u>

the higher is the number of the impacted quantum for the equal time counts of rest.	the less is the *time count of rest* at equal number of impacted quantum.

= <u>the higher is the intensity potential, the higher is the *frequency* of the impact.</u>

And obviously: $dmc_x^2 = dn_x$, which gives: $c_x^2 = \dfrac{dn_x}{dm};$ or in a more visual format: $dm = \dfrac{dn_x}{c_x^2};$ 1A5

"dm" in its conventional understanding means the change (of mass) in general.

“dm” here in 1A5 is also about the change of the number of the impacted quantum, related to the value of the quantum speed: **the higher the quantum speed is, the higher is the number of the quantum impacted**.

Ref. 1A2-1A4 1A6

With reference to the earlier $dt_{ox} = \frac{dn_x}{e_x}$; and $e_x = \frac{dmc_x^2}{dt_{ox}}$;

As $mc_x^2 \approx n_x$, or $n_x = f(c) \cdot m$, it seems the higher the quantum speed is, the shorter is the “time count of rest” at the inflexion. [$f(c)$ means: *function* of c]

! But $\lim \Delta t = 0$ is one and the same, “the shortest” for any value of the quantum speed! With reference to 1A4 (α), if it is about any frequency increase, it is not about the shortening of the shortest time count, rather about the increase of the quantum impact for the time count of $\lim \Delta t = 0$!

The intensity of the proton process after the inflexion is changing.

1A7

The actual intensity of the expansion is expressed by the actual time count of the developing impact: $dt_v = \frac{dt_o}{\sqrt{1 - \frac{v^2}{c_x^2}}}$;

This is for the utilisation of the intensity potential of the inflexion – for *sphere symmetrical expanding acceleration.*

The original impact remains the same, while the time count is increasing!

1A8 Ref. 1A1

The proton processes start at the anti-proton/proton *inflexion* with speed value $v = 0$, as the elementary process continues: $dt_{start} = \frac{dt_o}{\sqrt{1 - \frac{v^2}{c^2}}} = dt_o$;

$v = 0$ means no event and 1A1 above would mean no time $\Delta t = 0$ (duration).

Therefore if we want to formulate and calculate the highest intensity, still belonging to an event, it must be $\lim v = 0$ indeed, the closest event to and from the *inflexion.*

Ref. 1A2-1A8

The proton process is repeating the intensity of the anti-proton collapse and impacting the same equal number of quantum during the expansion. The intensities of the start however – from elementary process to elementary process – vary, while the shortest time count (of rest) remains always one and the same: $\lim \Delta t = 0$. **Therefore,** with reference to the explanations above, **the *quantum speed* values of the communication of the process at the *inflexions vary*** from elementary process to elementary process as well**!**

The higher the intensity/energy of the inflexion is, the higher is the quantum speed, and higher is the *frequency* of the starting intensity impact of the proton process.

The value of e_x in 1A2 is expressing the whole energy intensity potential of the inflexion.

1B1

The equation in 1A1 for the energy intensity of the proton process shall be written as: $e_p = \frac{dmc_x^2}{dt_o}\left(1 - \sqrt{1 - \frac{v^2}{c_x^2}}\right)$;

where dt_o is the time count of rest, expressing this status for all proton processes.

The c_x quantum speed in the formula specifies the intensity capacity of the process.

The intensity of the expansion 1B1 is decreasing during the proton process: the number of the impacted quantum of the expansion remains the same, but the time count of the impact – with the increase of *v*, the speed value of the expansion – is increasing.

The time count of the *electron process* is constant. Time dt_i of this expanded status corresponds to $\lim i_x = c_x$ value. But in this case the intensity capacity (energy) of the process, generated at the anti-proton/proton *inflexion* is the one which expires as drive:

The number of quantum, impacted at the inflexion is handing over the impact.

$$w_e = \frac{dmc_x^2}{dt_i\varepsilon_x}\left(1-\sqrt{1-\frac{(c_x-i_x)^2}{c_x^2}}\right);$$ with the time count of the process is infinite long. 1B2

The intensity of the anti-proton/proton inflexion $e_x = \dfrac{dmc_x^2}{dt_{ox}}$; results in $e_x = n_x w_e$; which gives $n_x = \dfrac{e_x}{w_e}$; 1B3

and $$n_x = \frac{dt_i\varepsilon_x}{dt_{ox}\left(1-\sqrt{1-\frac{(c_x-i_x)^2}{c_x^2}}\right)};$$ indeed. The time relation gives the number of the impacted quantum. 1B4

Ref. 1A5

With reference to 1A5 and 1B3: $dm = \dfrac{dn_x}{c_x^2}$; $dmc_x^2 = \dfrac{de_x}{w_e}$; or $w_e = \dfrac{de_x}{dmc_x^2}$ 1B5

The electron process, reaching the fully expanded status of the quantum impact of the anti-proton/proton inflexion *collapses* as *neutron process* under the driving impact of the electron process of the next cycle, it. The source of the drive, with reference to 1B2 above, is the energy potential of the expanded status – in fact the n_x number of the impacted quantum at the anti-proton/proton inflexion. The decreasing quantum impact of the electron process status means, the number of the impacted quantum (the potential of the expanded status) is decreasing. The cumulative quantum impact of the collapse in parallel is increasing in the neutron process. The neutron process is the accumulation of the quantum impacts of the expiring electron processes. The intensity of the collapse is increasing, the time count of the neutron process is shortening.

The formula of the collapse is:

$$\frac{dmc_x^2}{dt_i}\sqrt{1-\frac{(c_x-i_x)^2}{c_x^2}} - \frac{dmc_x^2}{dt_i\sqrt{1-\frac{i_x^2}{c_x^2}}}\sqrt{1-\frac{(c_x-i_x)^2}{c_x^2}} = \frac{dmc_x^2}{dt_i}\sqrt{1-\frac{(c_x-i_x)^2}{c_x^2}} - \frac{dmc_x^2}{dt_o}\sqrt{1-\frac{(c_x-i_x)^2}{c_x^2}};$$ 1B6

The drive has its limit. The entropy product of the process is the *quantum impulse*.
The quantum speed of the neutron/anti-neutron expansion this way is less – by the formulation of the *quantum* – than as it was at the anti-proton/proton inflexion.

The process is reformulating from expansion to the collapse. The time count is decreasing and the intensity of the collapse approaching the neutron/anti-neutron inflexion, which is: $\dfrac{dmc_x^2}{dt_n}\sqrt{1-\dfrac{(c_x-i_x)^2}{c_x^2}}$; 1B7

$\lim dt_{on} = \lim dt_{op} = \lim dt_o = 0$ the comment is in S.1.2

Quantum speed of $c_x\sqrt[2]{1-\dfrac{(c_x-i_x)^2}{c_x^2}}$ at the neutron/anti-neutron *inflexion* determines the number of its quantum impact potential. 1B8

The anti-neutron expansion repeats the intensity of the neutron collapse. The expansion is natural need. The accumulated intensity of the neutron collapse shall be utilised at the inflexion, impacting certain number of quantum during the anti-neutron expansion.

1B9 The intensities of the anti-electron process quantum drives of all elementary processes are equal: $IQ_{x-} = \frac{c_x^2}{\varepsilon_{x-}} = const$

and are impacting equal number of quantum during the anti-proton collapse.
The collapse of the anti-proton process – because of the generating quantum impulse on the anti-side as well – starts from an even more decreased intensity status, which is:

1B10 (The collapse goes similar way as to the one in 1B8, just the starting intensity is different.) $\frac{dmc_x^2}{dt_i}\sqrt{1-\frac{(c_x-i_x)^2}{c_x^2}}\sqrt{1-\frac{(c_x-i_x)^2}{c_x^2}}$;

1B11 The quantum speed value approaching the anti-proton/proton inflexion, coming from 1B10 is: $c_x\sqrt{1-\frac{(c_x-i_x)^2}{c_x^2}} = c_{x+1}$;

This is the quantum speed value c_{x+1} of the next cycle!

1B11 is the quantum speed value, where the new elementary process starts from. 1B11 gives the reason and the proof of the change. The time count of rest of the new proton process, following the inflexion is the same: $\lim \Delta t_o = 0$!

The quantum speed characterises the intensity of the inflexion and it is the speed value of the quantum communication of the elementary process. With reference to 1B11, each new elementary cycle starts with the modified quantum speed value of the new space-time.

The origin of the speed value of the quantum communication is the generation of the plasma, the collapse of the all over accumulated *Hydrogen* processes. All following processes repeat the cycle, with quantum speed values decreasing step by step.

Ref. 1A5 1A6

The number of the impacted quantum from cycle to cycle is reducing as 1B3 and 1B10 are demonstrating it. With reference to 1A5 and 1A6, this is the consequence of the change of the quantum speed value! [While the original quantum impact potential of the start (dmc_x^2) remains *quasi* the same in the formula. Quasi, because of the losses for the generation of the quantum impulses in each elementary cycle.]

S. 1.1

1.1 The meaning of the quantum speed

The value of the quantum speed is expressing the intensity of the event. This is the speed of the quantum impact of the collapse/expansion at the *inflexion*.
This is the reason the quantum speed of the plasma is of infinite high value: The quantum speed of the collapse of the *Hydrogen process* is establishing the intensity of the *plasma inflexion*! The impact of the plasma inflexions propagates with $\lim c_{pl} = \infty$ speed value.

The intensity of the impact will be lost step by step, through the *expansion* of the proton and the anti-neutron processes. But in these cases, as explained, the loss is about the frequency of the impact rather than the number of the impacted quantum.

In other words: the frequency of the impact depends on the actual status of the expansion/collapse (v), impacting different number of quantum or accumulating different number of quantum impacts for the unit period of time.

$E = mc^2$ is the total energy, generating in the *Hydrogen process/plasma inflexion.* 1C1

The event happens with $\frac{dmc_{pl}^2}{dt_{opl}} = f_{pl}$ intensity, and $\lim f_{pl} = \propto$. 1C2

With reference to 1A1-1A7, the expression in 1C1 corresponds to infinite large number of impacted quantum. And the event is not about losing on the impact, rather "losing" on the *frequency* = the increase of the time count. The reason is that the event of the expansion cannot lose on the $E = mc^2$ value, the total number of the impacted quantum!

For having any loss, work is necessary. But there is no work here. The proton process is energy transfer, with the "time is ticking on". *The loss of the frequency is obvious, since approaching the equal number of quantum, as it was at the start needs more time in the increasing space, result of the expansion.*

$$e_x = \frac{dmc_x^2}{dt_{pox}}\left(1 - \sqrt{1 - \frac{v^2}{c_x^2}}\right);$$ 1C3

This is energy transfer in time, where the intensity of the initial total number of the quantum impact of the expansion is losing on its frequency. v in 1C3 is representing the intensity of the impact. v is a relativistic figure, since the higher the quantum speed is, the less is the time count (higher the frequency) of the actual impact. Quantum speed values establish space-times. The transfer happens at this speed value of communication and the electron and anti-electron processes operate (with reference to the previous section) at the speed value of quantum communication.

With reference to 1B2 and 1B3 the electron and the anti-electron processes mean quantum impacts, until the work intensity is reaching the *quantum impulse* stage, while the time count (system) remains unchanged.

$$\frac{dmc_x^2}{dt_i \varepsilon_x}\left(1 - \sqrt{1 - \frac{(c_x - i_x)^2}{c_x^2}}\right);$$ Ref. 1B2 1B3

The intensity of the electron process is the end stage of the proton process: $\frac{dmc_x^2}{dt_i}$; with $E_{\inf} = mc_x^2$, the total energy of the *inflexion* remaining the same; 1C4

expiring this total energy by (as result of the drive) $\Delta E_{\exp} = mc_x^2 \sqrt{1 - \frac{(c_x - i_x)^2}{c_x^2}}$; in line with ε_x (and dt_i) of the elementary process. 1C5

With reference to the above and to 1B2 of the earlier section, there are two options for the reduction of the intensity of the quantum drive at constant $dt_i = const$ of the time system:

(1) either c_x, the speed of the quantum communication is changing,

(2) or m – in fact the number of the quantum impacted is changing,

(3) or both.

The answer is coming from the fact, that the electron and the anti-electron processes are not just about the reduction of the quantum impact. With reference to 1A5, this is the reduction of the number of quantum, originally impacted at the inflexions.

Ref. 1A5 1C6

It can be written as $\frac{dn}{dt_i\varepsilon_x}\left(1-\sqrt{1-\frac{(c_x-i_x)^2}{c_x^2}}\right)$; where $dm=\frac{dn_x}{c_x^2}$;

(With $n_x = const$ the quantum speed value should be the one to increase.)

The quantum speed is result of the anti-proton/proton inflexion, the characteristic of the intensity of the inflexion, establishing the number of quantum, impacted and the space-time of the elementary process, the fundament of the communication of the elementary process. The quantum impact cannot change the quantum speed value.
With reference to 1C6 above, the reduction of the intensity of the quantum drive does not influence the quantum speed of the energy transfer between the expansion and the collapse. The *blue shift* quantum drives of the electron and the anti-electron processes are acting at constant intensities, until they are capable at all to drive the process, until the intensity of the impact in 1B3 reaches the intensity value of the *quantum impulse*, *quantum*, not capable for drive. [As the drive is always coming from the next elementary cycle, the intensity of the drives has negative gradient.]

The quantum speed of the energy transfer from expansion to collapse remains of the same value. The only change in the energy transfer is the reformulation of the energy of the expansion. It will be less by a single quantum at the inflexion of the collapse than it was at the inflexion of the expansion:

Ref. 1B9 1C7

$$E_{n+1} < E_n \text{, meaning: } mc^2_{(n+1)} < mc^2_n \text{ and } c_{(n+1)} < c_n$$

Once the drive is over, the inflexion happens – at reduced quantum speed value, reduced space-time and as consequence at reduced intensity (frequency)!

Translating the situation to our circumstances on the *Earth* surface:
The acting quantum speed on the *Earth* surface is result of the sphere symmetrical expanding acceleration of the *Earth*. This is given to us, as the only quantum tool to use it. (The *Sunshine* is of higher quantum speed value. It slows down, as result of the conflict with the quantum impact of gravitation and of the electron process quantum impacts of the elementary processes of the atmosphere above the *Earth* surface.)

Quantum impacts, generated by our technical instruments, represent the characteristics of the elementary process of their generation. Light signals for example lose on their intensity, on the way approaching us.

The technical impact is: The quantum impact of gravitation (*qig*) is:

1C8

$$e_{ti}=\frac{dmc_x^2}{dt_i\varepsilon_x}\left(1-\sqrt{1-\frac{(c_x-i_x)^2}{c_x^2}}\right); \qquad e_{qig}=\frac{dmc^2_{Earth}}{dt_i(\varepsilon_{Earth}=1)}\left(1-\sqrt{1-\frac{(c_x-i_x)^2}{c_x^2}}\right);$$

All technical quantum impacts, including light signals mean work. Signals, approaching us are losing on their intensities and finally disappear. The propagation of technical signals is in conflict with the quantum impact of *gravitation* and with the elementary processes of our space-time. The quantum impact of gravitation (*qig*) is losing on its intensity the same way, but the source is infinite large, therefore it always corresponds to the demand – keeping the continuity of the quantum impact of gravitation at its constant intensity.

Having a technical *blue shift* impact signal within a vacuum chamber, the signal has its conflict with the quantum impact of gravitation, quantum speed of $c_{Earth} = 299792$ km/sec on its way of propagation. The relation of the capacities, as energy source, of the signals and *Earth* gravitation is so different, with the prevailing dominance of gravitation, that the measurement of the propagation of the signal (for example light) at certain distance will always give the quantum speed of *Earth* gravitation.
Why?
Because the condition for the propagation of any technical signals (impact) is: $e_{ti} > e_{qig}$.
The time systems, dt_i of the *quantum impact of gravitation* and of the technical signals are the same, and technical signals can only propagate within the space-time of the vacuum chamber, if $IQ_{ti} \geq IQ_{qig}$.
$IQ_{ti} > IQ_{qig}$ at the same time means conflict. The higher the impact is, the higher is the conflict. The source of the technical signal is in permanent demand for compensating the loss.
If the intensity of the signal is higher than the intensity of the quantum impact of gravitation – the conflict is obvious. If it less, the signal disappears. Signals propagate if their quantum speed is equal or more than the quantum speed of the quantum impact of gravitation.
Why?
Because gravitation keeps quantum within the vacuum chamber in a certain vibration, at the frequency of the quantum impact of gravitation. The propagation of signals is possible if the intensity of the impact, the speed of the signal corresponds to this "background" frequency or higher. Otherwise signals disappear in the conflict.
In the case of conflict, the conflict itself (in fact the light signal) is propagating with the quantum speed of the quantum impact of gravitation.
Why?
Because the propagating signal is *blue shift* impact itself as well, work intensity, which is approaching certain number of quantum for the unit period of time within the time system – one and the same for any kind of *blue shift* signals, including the quantum impact of gravitation (*qig*) – of the $\lim i = c$ speed value.
Because gravitation always provides the necessary number of quantum impacts in order to keep the continuity and the constancy of the *quantum impact of gravitation*, gravitation cannot locally disappear as result of the conflict.
Gravitation withstands the impact, generating the conflict:

$$n\frac{dmc^2_{Earth}}{dt_i\varepsilon_{Earth-}} = \frac{dmc^2_x}{dt_i\varepsilon_x}; \qquad \text{1C9}$$

The propagation of the signal means conflicts happen continuously, making by that the signal seen for us. The motion of the *blue shift* signal has been in practice continuously "stopped" by the quantum impact of gravitation. This stop is no other than meeting the propagating signal with increased number of *qig* and forcing it to fight.

The experienced light signal is the intensity/energy surplus, generated by the conflict, the consequence of the reduction of the intensity of the signal by *qig*.

Since the *IQ* value of the propagating signal is constantly reduced to the *IQ* value of gravitation, this is no other than the propagation of the signal by the quantum speed of gravitation on the *Earth* surface. While the propagation is driven by the source of the impact, the conflict has been managed. Once the energy/intensity supply from the origin is not capable to create the conflict any more, the signal (light) disappears.

S. 1.1.1

1.1.1. What the *quantum impact of gravitation* (*qig*) is about in spaces above certain structures, within closed premises and in the vacuum chamber?

The quantum impact of gravitation is impacting all elementary processes and elementary compositions above the *Earth* surface. The natural need/purpose is to ensure the continuity of the impact everywhere equally above the *Earth* surface.

1C10 $$IQ_{x-} = \frac{c_x^2}{\varepsilon_{x-}} = const$$

But the IQ_{x-} drives of the anti-electron processes of all elementary processes are equal to the IQ_{qig} drive of gravitation.

Therefore elementary processes do not have conflict with the quantum impact of gravitation!
It is taken and released without any impact to the elementary process, similarly as it happens in elementary communication and elementary evolution within the *Earth*.

The quantum impact of gravitation has not been stopped at the boundaries of solid or liquid subjects, as it has not been stopped in the case of gaseous aggregate states as well. The impact is propagating as the feeding source of the *quantum impact of gravitation* is infinite large. The quantum impact of gravitation is present everywhere (and in the vacuum chamber as well) until the intensity source of gravitation makes it possible.

S. 1.2

1.2
Additional arguments, explaining the variety of the quantum speed values in elementary processes

The intensities of the proton and the neutron processes in elementary communication are different. The relation of these two intensities is the one, making the difference.

1D1 $$\frac{dmc_x^2}{dt_{px}}\left(1-\sqrt{1-\frac{i_x^2}{c_x^2}}\right) \neq \xi \frac{dmc_x^2}{dt_{nx}}\sqrt{1-\frac{(c_x-i_x)^2}{c_x^2}}\left(1-\sqrt{1-\frac{i_x^2}{c_x^2}}\right);$$

Here in 1D1 above $dt_{px} \neq dt_{nx}$, as the drive of the electron process is increasing the intensity of the collapse:

1D2 $$dt_{px} = \frac{dt_o}{\varepsilon_p}; \text{ and } dt_{nx} = \frac{dt_o}{\varepsilon_n};$$

the relation of the two gives the intensity of the electron process, initiated by the quantum membrane, generated on the anti-side by the anti-electron processes.

This is not about a shorter time count than $\lim dt_o = 0$ (the closest to the inflexion), rather the duration of the process. At $\lim v = 0$, $dt_{nx} = dt_{px} = \lim dt_o = 0$

The absolute balance of the neutron and the proton processes for proton and neutron process dominant elementary processes is different.

The balance for the elementary processes with neutron process dominance is:

$$\frac{dmc_x^2}{dt_{px}\varepsilon_{px}}\left(1-\sqrt{1-\frac{i_x^2}{c_x^2}}\right) = \xi\frac{dmc_x^2}{dt_{nx}\varepsilon_{nx}}\sqrt{1-\frac{(c_x-i_x)^2}{c_x^2}}\left(1-\sqrt{1-\frac{i_x^2}{c_x^2}}\right) \qquad \text{1D3}$$

In the case of proton process dominance the balance is formulating on the anti-side.

The intensity of the electron process is function of the proton and the neutron processes : $\varepsilon_{ex} = \frac{\varepsilon_{px}}{\varepsilon_{nx}}\xi\sqrt{1-\frac{(c_x-i_x)^2}{c_x^2}} = \frac{dt_{nx}}{dt_{px}}$; 1D4

and the electron process drive is one of the specifics of the elementary process: $e_{ex} = \frac{dmc_x^2}{dt_i\varepsilon_{ex}}\left(1-\sqrt{1-\frac{(c_x-i_x)^2}{c_x^2}}\right)$; 1D5

The intensity of the anti-electron process, the drive of the anti-proton process is: $e_{ex-} = \frac{dmc_x^2}{dt_i\varepsilon_{ex-}}\left(1-\sqrt{1-\frac{(c_x-i_x)^2}{c_x^2}}\right)$; 1D6

Inflexions only change the direction of the process, the intensities are:

$\varepsilon_{nx} = \varepsilon_{nx-}$; and $\varepsilon_{px} = \varepsilon_{px-}$; in line with this: $\varepsilon_{ex} = \frac{1}{\varepsilon_{ex-}}$. 1D7

The neutron and the anti-proton processes are both collapses, driven by the electron and the anti-electron processes. The intensities of the expanding acceleration of the anti-neutron and the proton processes are results of the intensity potential, accumulating during the collapse before the inflexion.

$$\frac{dmc_x^2}{dt_{nx-}}\sqrt{1-\frac{(c_x-i_x)^2}{c_x^2}}\left(1-\sqrt{1-\frac{i_x^2}{c_x^2}}\right) = \frac{dmc_x^2}{dt_{nx}}\sqrt{1-\frac{(c_x-i_x)^2}{c_x^2}}\left(1-\sqrt{1-\frac{i_x^2}{c_x^2}}\right);\ \text{and} \qquad \text{1E1}$$

$$\frac{dmc_x^2}{dt_{px-}}\sqrt[2]{1-\frac{(c_x-i_x)^2}{c_x^2}}\left(1-\sqrt{1-\frac{i_x^2}{c_x^2}}\right) = \frac{dmc_x^2}{dt_{px}}\sqrt[2]{1-\frac{(c_x-i_x)^2}{c_x^2}}\left(1-\sqrt{1-\frac{i_x^2}{c_x^2}}\right); \qquad \text{1E2}$$

[The basics of the time counts of the anti-neutron and the anti-proton processes are similar to the earlier explanations. But more discussions about this subject will be in Section 2.]

where in 1E1 and 1E2 $c_{x+1} = c_x\sqrt{1-\frac{(c_x-i_x)^2}{c_x^2}}$ 1E3

In simple terms it means: $|+N| = |-N|$; and $|+P| = |-P|$; but $N \neq P$; and $-N \neq -P$ 1E4

The inflexions are repeating the intensities just in the opposite direction.

This means, the intensity of the expansion is equal to the accumulating intensity of the collapse, driven by the electron and the anti-electron process drives. The durations of the collapses are equal to the durations of the electron or the anti-electron processes. There is no collapse without drive. The drives are the ones making the difference.

The neutron processes are neutral and might be driven by electron process drives of other elementary process. This also determines the intensity relations of the anti-direction. Anti-processes do not communicate. Anti-processes are the instruments of the elementary control, leading back the elementary process to the original proton process start.

1F1
$$\frac{dmc_x^2}{dt_i\varepsilon_{ex-}}\left(1-\sqrt{1-\frac{(c_x-i_x)^2}{c_x^2}}\right) \neq \frac{dmc_x^2}{dt_i\varepsilon_{ex}}\left(1-\sqrt{1-\frac{(c_x-i_x)^2}{c_x^2}}\right);$$

The proton process has its certain intensity value, which corresponds to the accumulated intensity value of the anti-proton collapse, which is driven.

$\varepsilon_{ex} > 1$ on the direct side of the elementary process means proton process dominance and $\varepsilon_{ex-} < 1$ on the anti-direction;

$\varepsilon_{ex} < 1$ on the direct side means neutron process dominance and $\varepsilon_{ex-} > 1$ on the anti-direction.

As the quantum speed value within the elementary process is quasi one and the same

1F2	having $\varepsilon_{ex} > 1$ on the direct side, the intensity of the quantum drive on the anti-side shall be of increased value, since the dominance of the proton process on the direct side can only be guaranteed in this case. Therefore $\varepsilon_{ex-} < 1$ and $IQ_{ex-} > IQ_{ex}$.	$\frac{c_x^2}{\varepsilon_{ex-}} > \frac{c_x^2}{\varepsilon_{ex}};$
1F3	having $\varepsilon_{ex} < 1$ on the direct side and having this way neutron process dominance, the case is the opposite as $\varepsilon_{ex-} > 1$ on the anti-side and $IQ_{ex-} < IQ_{ex}$.	$\frac{c_x^2}{\varepsilon_{ex-}} < \frac{c_x^2}{\varepsilon_{ex}};$

1F4 $\varepsilon_{ex} > 1$; and $\varepsilon_{ex-} < 1$; means proton process dominance:

1F5 $f = (\varepsilon_{ex} - 1)\frac{dmc_x^2}{dt_i\varepsilon_{ex}}$; is the drive of the electron process surplus,

1F6 $\varepsilon_{ex} < 1$; and $\varepsilon_{ex-} > 1$; - neutron process dominance

1F7 $g = (\varepsilon_{ex-} - 1)\frac{dmc_x^2}{dt_i\varepsilon_{ex-}}$; is the quantum impact of gravitation,

If c_x the quantum speed would be a value, constant for all elementary processes,

- first of all the proton processes would not be of the experienced measured stability, similar events for all elementary space-times;
- it could happen, that *the intensity of the proton process* of a neutron process dominant elementary process with high periodic number, (as for example the *Uranium* process), would be only portion of the intensity of the proton process of a proton process dominant elementary process, (as for example the *Oxygen* process), with low periodic number.

Ref. Tabl. 3.1 The value of the intensity coefficient of the anti-electron process of the *Uranium* process is *1.5679*, while the same for the *Oxygen* process is *0.9847*.

With reference to the above, the intensities of the drives of the anti-proton processes would be:

for the *Uranium* process: $\frac{c^2_{const}}{\varepsilon_{Ue-}} = \frac{c^2_{const}}{1.5679}$, and for the *Oxygen* process: $\frac{c^2_{const}}{\varepsilon_{Oe-}} = \frac{c^2_{const}}{0.9847}$ 1F8

The proton processes, as similar events for all space-times of all elementary processes require: $\frac{c^2_x}{\varepsilon_{x-}} = const$; 1F9

and in the contrary to this this, with reference to 0I8: $\frac{c^2_{const}}{1.5679} \neq \frac{c^2_{const}}{0.9847}$; 1F10

In the case the speed of the quantum communication would be a constant value for all elementary processes and space-times, as principal point, there would be no return from the *plasma* process and the elementary evolution could not happen. With the increase of the value of the intensity coefficient of the anti-electron process, the intensity of the proton process would be decreasing.

The plasma state is with $\lim \varepsilon_{pl-} = \infty$, as the intensity of the collapse is of infinite high value: $\lim \varepsilon_{pl} = 0$.

It would mean in fact no quantum drive on the anti-side: $\frac{c^2_{const}}{\lim \varepsilon_{pl-} = \infty} = 0$; 1G1

It would be similar to the direct process of the *Hydrogen* process with $\lim \varepsilon_H = \infty$ without end, as $\frac{c^2_{const}}{\lim \varepsilon_H = \infty} = 0$; 1G2

While with $c_{pl} = \infty$, as the real case is:

the quantum speed is changing, growing or decreasing in line with the change of the intensity coefficient of the anti-process. $\frac{c^2_x}{\varepsilon_{ex-}} = const$ 1G3

which also means: $c^2_x \cdot \varepsilon_{ex} = const$ 1G4

With the increase of the value of the intensity coefficient of the anti-electron process the quantum speed is increasing in parallel as well. In the false case of $c_x = const$, the intensities of the anti-electron process IQ_{x-} quantum drives of the anti-proton processes would be different for all elementary processes!

While for the stability of the elementary evolution it should be of constant value: $e_{ex-} = \frac{dmc^2_x}{dt_i \varepsilon_{ex-}} \left(1 - \sqrt{1 - \frac{(c_x - i_x)^2}{c^2_x}}\right) = const$ 1G5

The intensities of the IQ_{x-} quantum drives are equal and the proton processes as events are similar, but happening within different space-times, with different quantum speed values and impacting different number of quantum. Therefore the value of the speed of the quantum communication of the elementary processes is of infinite large variety. The change reflects the variety of the relation of the intensities of the proton and neutron processes!

Neutron process dominance generates anti-electron process surplus on the anti-side, proton process dominance generates electron process surplus, but on the direct side.
The continuity cannot be disrupted. Neither in the case of the overload of the drives, nor in the case of the lack of the sufficient drives of the process.
The overload is always given off. The anti-electron process overload at the anti-side establishes the quantum impact of *gravitation*, the electron process surplus at the direct side the quantum impact of communication.
The similarity in the proton processes as events is a kind of additional proof to the *quantum impact* and the *mechanical impact* of *gravitation.*

S. 1.3

1.3
The elementary processes are about *change*

The *inflexion* is the point where the whole energy/intensity capacity of the process is generating. The collapse, the *increase of the intensity* means in conventional terms the increase of the frequency. The expansion, the *release of the accumulated intensity capacity* means the decrease of the frequency. The quantum speed, the characteristic of the inflexion (of "*the change*") establishes the space-time of the impact. Meaning: the intensity of the change at the *inflexion* has its energy coverage, or quantum impact, or information of certain frequency propagating – about the change.
Once a collapse of specific intensity reaches the point of the highest frequency, the highest intensity of its compression, it turns back at the inflexion and the expansion starts. The expansion is no other than the self-utilisation of the full collapse. The quantum speed is the speed of the transportation of the energy/intensity from the expansion to the collapse.

In the case of the *plasma,* the quantum impact, leaving the inflexion is of infinite high intensity. The impact is propagating for infinite short time and at infinite long distance. In the contrary, in the *Hydrogen* process the quantum impact of the inflexion is of infinite low intensity and propagating for infinite long time and at infinite short distance.
The proton process is the tool, transporting the energy/intensity, the information of certain frequency. The original intensity of the change, the frequency, the information, transported by the quantum speed is expiring in time and space. The meaning of the space in this case is no other than the number of quantum, approached.
The release of the impacted by the proton process quantum, the information – is giving over the intensity, generated at the inflexion for recreating the event. This is not just about the principle of *"there is no time without event and there is no evet without time"*, but about the balance of the space-time of the inflexion. There is a natural need for the recreation of the expansion of the inflexion, which needs collapse. But the collapse needs external drive.
The neutron/anti-neutron inflexion returns the process and the anti-process brings back the change to the start again.
The energy/information is losing on its intensity/frequency during the transport and ending with the *entropy* product of the elementary process: the *quantum impulse* (quantum).

The elementary process ***means change of the number of the impacted quantum in time.*** The changing time count from infinite long to infinite short and from infinite short to infinite long; the changing frequency from infinite high to infinite low and vice versa. The origin of the change is the *inflexion* – the event without time. The change of the number of the impacted quantum in time creates events, elementary processes, creates impacts, those which are part of the exchange and those quantum impulses (quantum), which never expire and compose the quantum space.

Elementary processes mean infinite large number of events happening in parallel.
Inflexions happen constantly, the generation of quantum impulses is continuous.

1.4
The *quantum impulse* (quantum) and the elementary balance

S.
1.4

With reference to 1A2 - 1A7, the time count of the quasi rest status is one and the same for all elementary processes, with the intensity characteristics of the proton process expansion and the neutron process collapse. Each elementary process has its own quantum speed value and the transformation of the energy happens within the space-time of the elementary process.

Ref 1A2 1A7

With reference to 1B11, the incoming quantum speed of the anti-proton/proton inflexion is establishing the speed of the quantum communication of the next elementary cycle, the speed of the transportation of the energy/information of the elementary process.

Ref. 1B11

The drives of the *collapses* are the electron and the anti-electron processes, operating in equal time systems. The collapse itself is the capacity increase of the process, the increase of the frequency of the impact. The *expansion*, in the contrary, means the utilisation of the accumulated intensity capacity of the collapse by the expanding acceleration from $v = 0$ of the inflexion to $v = i = \lim a \cdot \Delta t = c$ of the electron process.
The *proton process* starts at $\lim \Delta t_o = 0$ and ends by $\lim dt_i = \infty$, time counts, corresponding to $v = i = \lim a \cdot \Delta t = c$, the final speed value of the sphere symmetrical expanding acceleration.
The *electron process* is acceleration at constant dt_i time system, work with the expiration of the energy/intensity, the drive of the neutron collapse at the quantum speed of the communication. The time count of all electron processes is constant and $\lim dt_i = \infty$.

The electron process is a *blue shift* impact:

$$e_{drive} = \frac{dmc^2}{dt_i} - \frac{dmc^2}{dt_{fe}} = \frac{dmc^2}{dt_i} - \frac{dmc^2}{dt_i}\sqrt{1 - \frac{(c-i)^2}{c^2}} \quad \text{1H1}$$

The *neutron process* is sphere symmetrical accelerating collapse, driven by the electron process, from the fully expanded status of the drive-impacts at speed $v = i$ and time system of dt_i to the neutron/anti-neutron *inflexion* to $v = 0$ and $\Delta t_{\inf} = 0$.

Ref. 1B6 1H2

$$\frac{dmc^2}{dt_i}\sqrt{1 - \frac{(c-i)^2}{c^2}} - \frac{dmc^2}{dt_i\sqrt{1-(i^2/c^2)}}\sqrt{1 - \frac{(c-i)^2}{c^2}} = \frac{dmc^2}{dt_i}\sqrt{1 - \frac{(c-i)^2}{c^2}} - \frac{dmc^2}{dt_o}\sqrt{1 - \frac{(c-i)^2}{c^2}};$$

The principal points of the intensity value of the electron process are:

- the electron process *blue shift* drive is working out its intensity capacity, but cannot reach the fully expanded (*fe*) status, as its driving impact is limited:

1H3 $$e_e = \frac{dmc_x^2}{dt_i\varepsilon_x}\left(1-\sqrt{1-\frac{(c_x-i_x)^2}{c_x^2}}\right);$$

Meaning: the frequency of the impact cannot be *zero* and $\lim dt_{fe} = \infty$.

All electron processes contribute to the neutron collapse and incorporate their quantum impact but one, the last with not sufficient intensity. The neutron process starts from this not exactly fully expanded – quantum entropy – electron process status.

Ref. 1D2

With reference to 1D2, the equation of the elementary balance in absolute terms is:

1H4 $$\frac{dmc_x^2}{dt_p\varepsilon_p}\left(1-\sqrt{1-\frac{i^2}{c^2}}\right) = \xi\frac{dmc_x^2}{dt_n\varepsilon_n}\sqrt{1-\frac{(c-i)^2}{c^2}}\left(1-\sqrt{1-\frac{i^2}{c^2}}\right) \quad \text{where } \lim\xi > 1$$

The equation above is representing the generating intensity at the anti-proton/proton inflexion, which is transported to the neutron/anti-neutron inflexion. The intensity of the proton process also includes the intensity of the electron process and the generating, if any, electron process surplus as well!

The energy of the neutron process is the accumulating intensity of the collapse, also taking over the generating energy of the expansion from the proton process.

The absolute formula in 1H4 gives the balance in general independently on the intensity of the proton and the neutron processes.

Elementary processes have their own quantum speed values and own space-times. It also gives the absolute balance in the case of the electron process drives of the collapse of other, "external" neutron process/es: The neutron processes are neutral, the electron processes have their own certain intensity values and the proton processes always cover the intensity need of the neutron collapse, driven by the electron processes. The correspondence of the intensity relations has been controlled by the anti-processes.

In the case the electron process is in surplus and the *blue shift* impact is used for external purpose (as for the elementary processes with proton process dominance), the balance in 1H4 is not valid anymore. Electron processes are also acting as external quantum impacts; the intensity of the proton process includes the energy/intensity potential of the electron process quantum impact as well.

Ref. 1A3

Neutron processes are neutral and the proton processes of all elementary processes are similar events, just, with reference to 1A3, of infinite large variety of quantum speed and intensity values. This means the operating quantum speed values, and this way the space-times and the time counts of the elementary processes vary.

Ref. S.2

More about the specifics of the proton processes is in Section 2.

Ref. 1D2

dt_p and dt_n in 1H4 are the time counts of the proton and neutron processes, with reference to 1D2, with certain intensities of the change.

ε_x is the intensity relation of the proton and neutron processes.

1./

$$\frac{dt_n}{dt_p} = \xi \frac{\varepsilon_p}{\varepsilon_n}\sqrt{1-\frac{(c-i)^2}{c^2}}\ ; \text{ and } \frac{\Delta t_n}{\Delta t_p} \cong \varepsilon_e \xi\ ;$$

2./

$$\lim_{i=\lim a\Delta t=c} \xi = \lim \frac{dt_n \varepsilon_n}{dt_p \varepsilon_p} \frac{1}{\sqrt{1-\frac{(c-i)^2}{c^2}}} = 1$$

approaching it from above!

1I1

1I2

With reference to 1H4, there is a remaining *quantum energy* intensity portion "reserve" within the balance between the expansion and the collapse – the *entropy* consequence of the process. [and also: $dt_p \varepsilon_p = dt_n \varepsilon_n = 1$].

During the electron process the acceleration happens at constant speed, at constant time count. The operating intensity of the electron process expires as it is utilised as drive of the neutron collapse. The intensity of the proton process also includes the infinite low intensity value of the *remaining quantum entropy* as well.

The remaining and acting *quantum entropy* means, the electron process cannot run out to full expansion, reaching *zero* intensity. The neutron collapse this way always starts from a not fully expanded electron process status, while the intensity of the expansion has been generated in its full volume during the anti-proton/proton inflexion. There is always a remaining residual intensity impact within the electron process drive of infinite low intensity, the reason of the generation of the *quantum impulse*, *energy quantum* – (see next section):

Ref. S.1.5

$$\lim q = \lim \Delta_{quantum} = (1-\zeta)\frac{dmc^2}{dt_i \varepsilon_e}\left(1-\sqrt{1-\frac{(c-i)^2}{c^2}}\right) = 0; \qquad \text{where } \lim_{0\to 1} \zeta = 1$$

1I3

The missing electron process *blue shift* drive is the intensity of the *quantum impulse*!

There is here an important point to be noted:

> The generating electron process or anti-electron process *surplus*, either on the direct or on the anti-sides, includes the intensity value of the generating *quantum entropy*, the *quantum impulse* as well. There is no separate generation in this case of a remaining quantum entropy impact and the surplus.

This means:

- in the case of neutron process dominant elementary processes the *quantum impulse* is generating on the direct side, as the elementary balance is formulating on the direct side;
- in the case of the 8 proton process dominant elementary processes the *quantum impulse* is generating on the anti-side, as the balance is formulating on the indirect side.

The intensities of the direct and the anti-processes are reciprocal, independently on the generating electron, or anti-electron process surplus.

$\varepsilon_{n-} = \varepsilon_n$ and $\varepsilon_{p-} = \varepsilon_p$ 1I4

this way: $\varepsilon_{x-} = \frac{1}{\varepsilon_x}$; 1I5

The balance in the case of proton process dominance is:

$$\frac{dmc_x^2}{dt_{n-}\varepsilon_{n-}}\left(1-\sqrt{1-\frac{i^2}{c^2}}\right) = \zeta \frac{dmc_x^2}{dt_{p-}\varepsilon_{p-}}\sqrt{1-\frac{(c-i)^2}{c^2}}\left(1-\sqrt{1-\frac{i^2}{c^2}}\right);$$

1I6

- In the case of neutron process dominance, the anti-electron process surplus of equal *IQ* quantum drives generates the *quantum impact of gravitation.*
- The intensity/energy potential of the anti-neutron expansion, utilised as anti-proton process cover represents the energy/intensity of the elementary evolution. The remaining not utilised part of the anti-neutron expansion is the *mechanical impact of gravitation.* This is the acting, in the elementary evolution cooling process of the plasma, the increase of the time count and the decrease of the intensity.

S. 1.5

1.5 The generation of the *quantum impulse*

Diagram 1.1 below is about the generation of the *quantum impulse* in a neutron process dominant elementary process.

The intensities of the proton and the electron process expansion are generating at the anti-proton/proton inflexion in its whole value. The value of the proton process intensity is decreasing step by step. It always contains the intensity of the electron process as well.

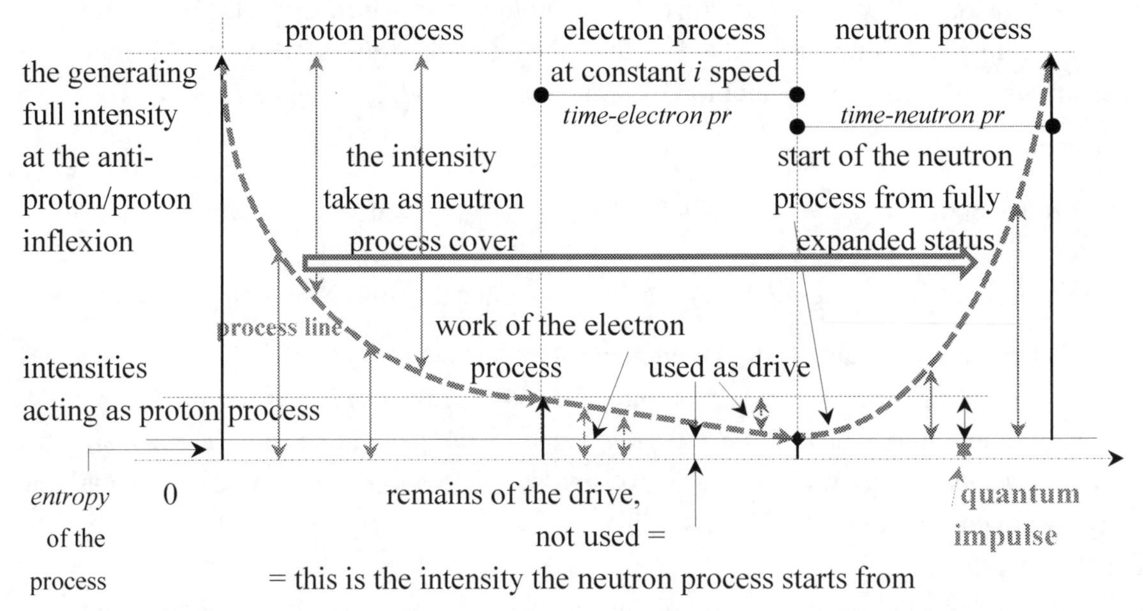

Diag. 1.1

Diagram 1.1

- the *red arrows* below the process line are the intensities of the proton process, still available, driving the expansion;
- the *blue arrows* above the process line are the ones have already been transmitted as intensity cover to the neutron process.

The intensity of the electron process is presented in its full value at the start.

The drive utilises the intensity of the electron process as *blue shift* impact, until it is capable to drive the collapse at constant $i = \lim a\Delta t = c$ speed.

The decreasing tendency of the work capacity of the electron process drive is presented by the red arrows in dash lines below the electron process line.

ε_x, the intensity coefficient of the electron process is the relation of the intensities of the proton and the neutron processes. This coefficient is constant and characterises the elementary process. With reference to 1D3, the value of the operating intensity depends on the capacity and the operating quantum drive of the electron process, characterised by the $(1-\zeta)$ multiplayer in 1I3. Ref. 1I3

The higher the value of the *IQ* quantum drive is, $IQ_x = \frac{c_x^2}{\varepsilon_x}$; 1I7

the higher is the gradient of the expiration of the intensity,

the more is the intensity of the elementary process and therefore the higher is the number of the operating in parallel elementary cycles.

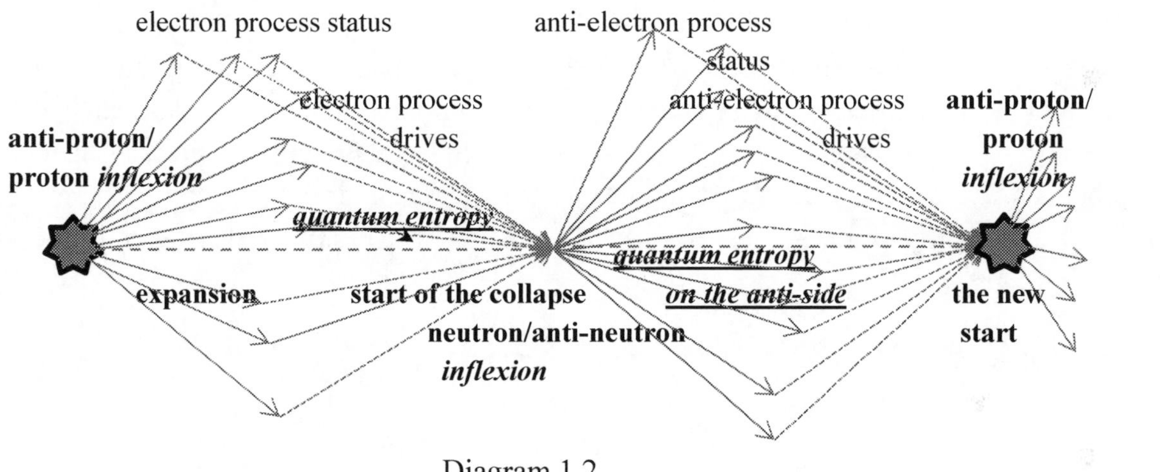

Diagram 1.2

Diag. 1.2

The electron process has its driving impact, until its capacity is capable to drive the collapse – reaching $\lim q = 0$. This is the start of the neutron process; the remaining intensity impact of the electron process, not capable any more to drive the collapse.

The end of the proton process is of certain frequency: $\frac{dmc_x^2}{dt_p}\sqrt{1-\frac{i_x^2}{c_x^2}} = \frac{dmc_x^2}{dt_i}$; 1J1

The frequency at the end of the electron process is: $\lim \frac{dmc_x^2}{dt_i \varepsilon_x}\sqrt{1-\frac{(c_x - i_x)^2}{c_x^2}} = 0$; 1J2

where the intensity coefficient in the nominator is not about the correction of the time system or the quantum speed value; it corresponds to the intensity relation of the proton and the neutron processes, established by the quantum membrane of the anti-electron processes.

The frequency, the intensity at the start of the neutron process obviously is: $\lim \frac{dmc_x^2}{dt_i}\sqrt{1-\frac{(c_x - i_x)^2}{c_x^2}} = 0$; 1J3

1J4 At the end of the slow-down, the frequency increase of the neutron process is: $$\frac{dmc_x^2}{dt_i\sqrt{1-(i_x^2/c_x^2)}}\sqrt{1-\frac{(c_x-i_x)^2}{c_x^2}}=\frac{dmc_x^2}{dt_n}\sqrt{1-\frac{(c_x-i_x)^2}{c_x^2}};$$

There is a "missing intensity within the cycle, value of:

1J5 $$\lim\frac{dmc_x^2}{dt_i\varepsilon_x}\left(1-\sqrt{1-\frac{(c_x-i_x)^2}{c_x^2}}\right)=0$$ corresponding to the missing frequency of the *quantum impulse* indeed!

S. 1.6

1.6 The common space-time of the proton and neutron processes

$e_p = e_{\mathrm{cov}} + e_{drive+}$; it is the energy, provided by the proton process;

$e_n = e_{drive-} + e_{\mathrm{cov}-}$; it is the energy, utilised by the neutron process;

1K1 The provided and the utilised drives at the two sides are different: $e_{drive+} \neq e_{drive-}$

The specificity of these processes further in 1K3 and 1K4 is that they are the equal parts of the same global process, just separated. The released cover (with +) on the expanding proton process side is initiated by the demand (with –) on the other, collapsing neutron process side. The receipt at one side is equal to the release on the other.

$v = 0$ at the proton process side means *inflexion*, dt_i at the side of the neutron process means time system of the electron process.

1K2 The change of $\frac{dmc_x^2}{dt_p}$ from the *inflexion* to dt_i is more

1K3 than the change from $\frac{dmc_x^2}{dt_i}\sqrt{1-\frac{(c_x-i_x)^2}{c_x^2}}$ to the $\frac{dmc_x^2}{dt_n}\sqrt{1-\frac{(c_x-i_x)^2}{c_x^2}}$ *inflexion*.

1K4 (1) The intensity, generated at the anti-proton/proton *inflexion* is a certain value, generated in its whole value. This means a certain starting intensity, frequency of:	$\frac{dmc_x^2}{dt_p}$;
1K5 (2) This frequency is in its natural decrease during the expansion (called as proton process) step by step. The reason of the decrease is to balancing the space-time, as there is another	$\frac{dmc_x^2}{dt_p}\left(1-\sqrt{1-\frac{v^2}{c_x^2}}\right)$;
1K6 process within the same space-time, which however is about the increase of the intensity – the neutron process.	$\frac{dmc_x^2}{dt_i\sqrt{1-(v^2/c^2)}}\sqrt{1-\frac{(c_x-i_x)^2}{c_x^2}}\left(\sqrt{1-\frac{v^2}{c_x^2}}-1\right)$;

The communication is without loss and happens at the speed of the quantum communication of the space-time of the elementary process. This *pulsation* is the process establishing not just the space-time, but also the event, which is establishing *time*!
From the equations is seen that $v \neq c_x$. There would be no communication at all in this case.
The case of $v = c_x$ are the *inflexions* on both sides, as earlier specified, for $\Delta t_{\mathrm{inf}} = 0$!

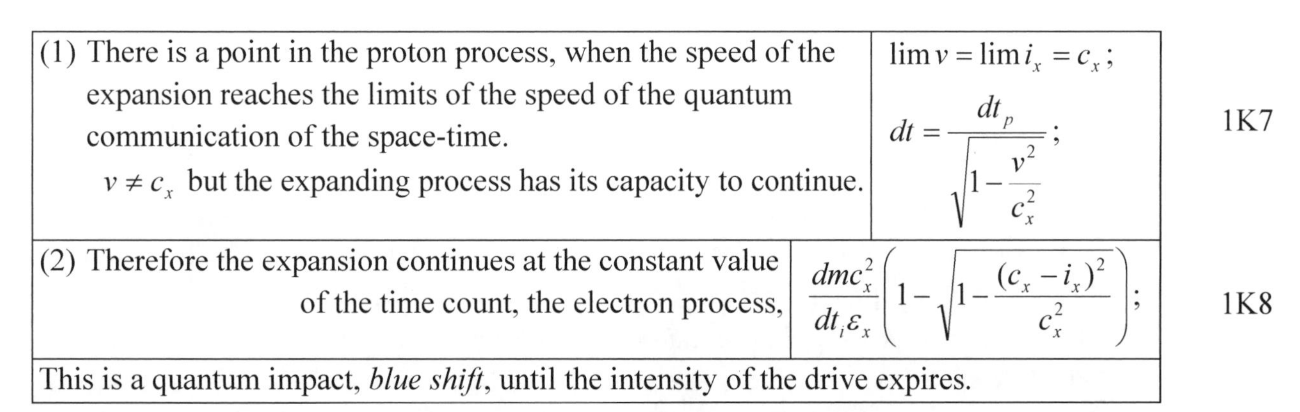

(1) There is a point in the proton process, when the speed of the expansion reaches the limits of the speed of the quantum communication of the space-time. $v \neq c_x$ but the expanding process has its capacity to continue.	$\lim v = \lim i_x = c_x$; $dt = \frac{dt_p}{\sqrt{1-\frac{v^2}{c_x^2}}}$;	1K7
(2) Therefore the expansion continues at the constant value of the time count, the electron process,	$\frac{dmc_x^2}{dt_i\varepsilon_x}\left(1-\sqrt{1-\frac{(c_x-i_x)^2}{c_x^2}}\right)$;	1K8
This is a quantum impact, *blue shift*, until the intensity of the drive expires.		

The electron process drives the neutron collapse, but its final intensity value cannot come down to zero.

The intensity capacity, generated in the anti-proton/proton inflexion cannot come to its full use. There is an intensity capacity of infinite small value, acting within the space-time.

p and ***n*** in Figure 1.1 characterise the capacities at the start of the expansion and at the end of the collapse.

$\boldsymbol{p - n} = \Delta q$ ***- the capacity portion, not used in the cycle!***

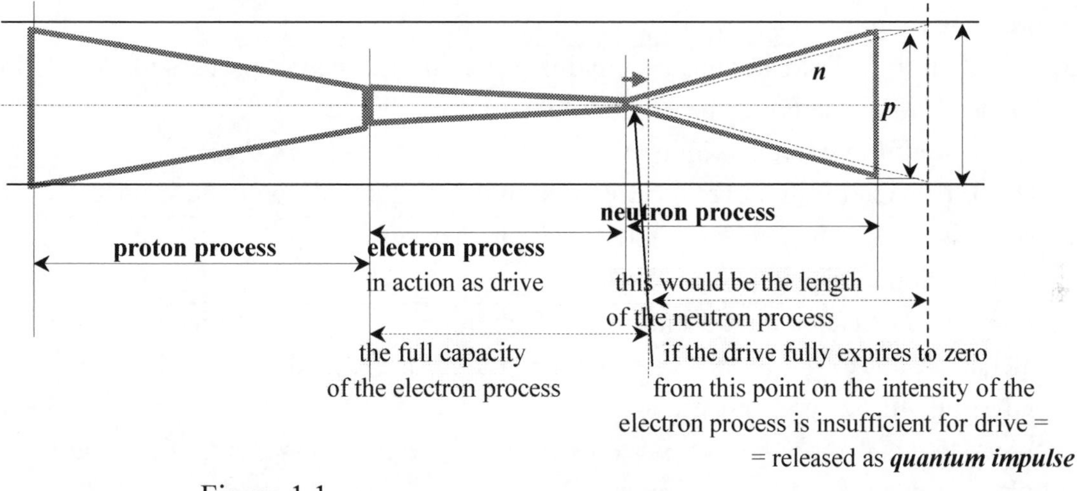

Figure 1.1

Fig. 1.1

p and ***n*** would only be equal if the electron process runs out to zero intensity. But in this case there would be no way for any collapse to be started.

There is an *intensity capacity of infinite small value, acting* within the space-time =

= the *quantum impulse*.

There are infinite high numbers of elementary processes acting as certain elementary process-composition, which are generating infinite large number of *quantum impulses*.

The duration of the elementary cycles within our space-time is a category of 10^{27} years.

The generation of the quantum impulse on the direct and the anti-direct sides of the processes is identical. In the case of surplus, the surplus includes the quantum impulse.

The lines in dash show the equality in this case. Figure 1.1 is in absolute values.

1K9

(3) This is the status, when the intensity of the expansion is $\frac{dmc_x^2}{dt_i}$;	with the time count of $dt_i = \frac{dt_p}{\sqrt{1-\frac{i_x^2}{c_x^2}}}$; demonstrating there is still intensity to expand.

- The electron process is crucial.
  Not just because of the workout of the remaining intensity of the expansion, but also for the initiation of the collapse!
- The constant drive toward the collapse is the demand of the space-time.
  Why? Because if v, the speed value of the collapse would not constantly grow in 1I3, there would be no demand, initiating the process!
- The *blue shift* quantum impact of the electron process generates this impact.
  Electron process *blue shift* impact drives the neutron process. The intensity/energy exchange goes on and the process continues, ensuring the pulsation (= life) in time.

(4) There will be a point, until the electron process is capable to drive the collapse – up to certain intensity: reaching $\lim v = 0$ by the collapse.
$v = 0$ means $\Delta t = 0$ and means: *inflexion*.

1K10

(5) Coming to the inflexion by the collapse, it is obvious, that with reference to 1I4, the intensity at the start in 1I2 could not be used in full.	$\frac{dmc_x^2}{dt_n}\sqrt{1-\frac{(c_x-i_x)^2}{c_x^2}}$;

(6) There will be a residual impact, remaining within the electron process of infinite low intensity, which could not be used as drive of the collapse.
This **remaining portion** within the drive is: ***the quantum impulse***!
The remains of the drive are the quantum impulses composing quantum space!

(7) There is important to make a note here that each electron processes drive is coming from the following elementary cycle.

Quantum impulses cannot and do not drive each other as they are of the smallest, infinite low intensity values. The *energy quantum* cannot be driven and cannot be in conflict, as they have no impact and are out of the elementary communication (with no electron process quantum drive, with no proton process cover). *Quantum impulses* transfer all quantum impacts without any modification!
Quantum impulses (quantum) are not about "moving" quantum impacts.
There is a difference between

- the quantum impacts initiated by the *blue shifts* of the electron and the anti-electron processes and
- the *quantum impulse*, the residual intensity of the electron and the anti-electron process drives.

➢ *Quantum impulses* (*quantum*) are not and cannot be in conflict – even having in parallel different quantum speed and intensity values at their generation; therefore
➢ the *quantum impulse* of infinite low intensity remains infinite low value.
➢ *Quantum impulses* are of infinite low intensity, independently on the original values of the quantum speed and the intensity of their generation.
➢ They transfer all *blue shift* impacts without any modification.

The *quantum impulse* (quantum) is the least possible quantum impact.

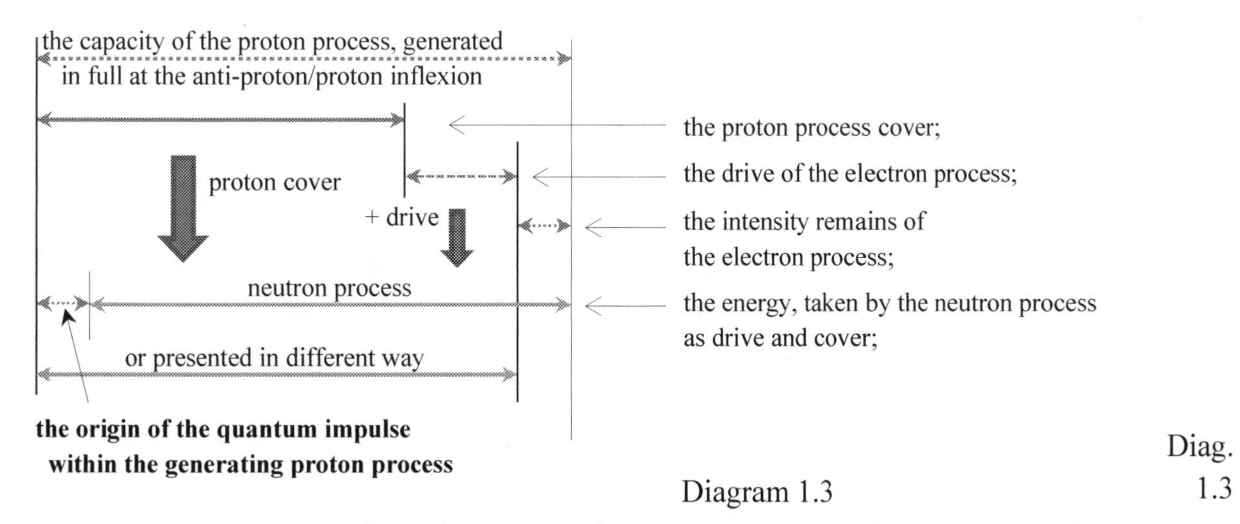

Diagram 1.3

Diag. 1.3

Diagram 1.3 is about the generation of the *quantum impulse* in neutron process dominant elementary process.

Diagram 1.4 below represents the generation of the *quantum impulse* in proton process dominant elementary process.

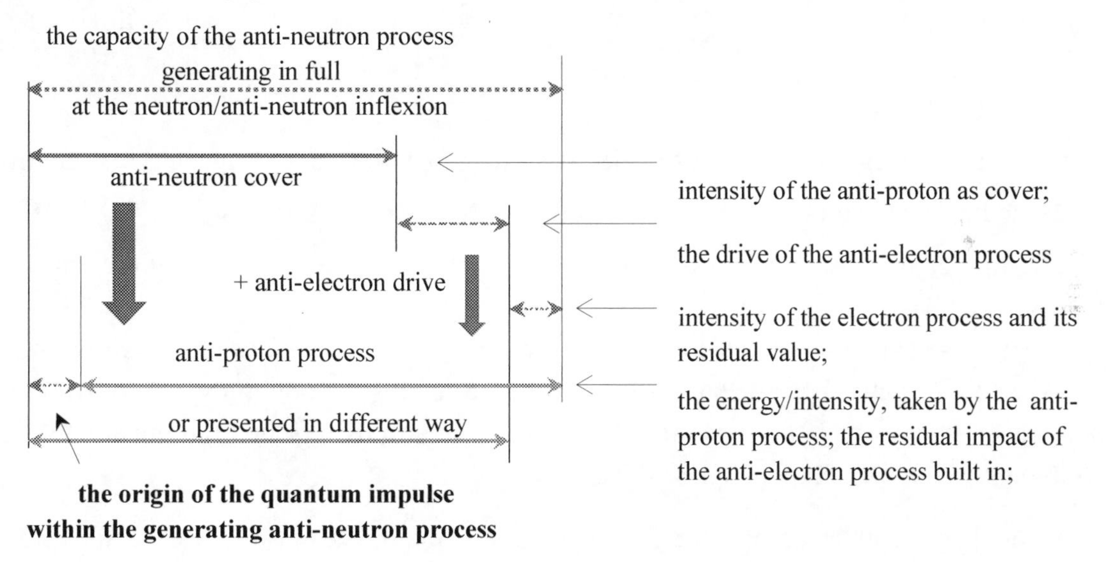

Diagram 1.4

Diag. 1.4

The anti-process is working similar way, just the capacity of the expansion is born at the neutron/anti-neutron inflexion and the collapse is the anti-proton process. The anti-process is establishing the proton process and generates the *quantum impact of gravitation.*

In the case of neutron process dominant elementary processes the *quantum impulse* is generating on the direct side.

In the case of proton dominant elementary processes the *quantum impulse* is generating on the anti-side.

There is no difference whether the generating *quantum impulse* was of electron process *blue shift* impact of high or low value *IQ* drive. There is no difference whether the source

of the generation of the *quantum impulse* was the *Uranium* process, the *Helium* process or any other processes. The meaning of $\lim q = 0$ is the same for all elementary processes.
But the ***Hydrogen* process does not generate quantum impulse,** as its elementary cycle has not been completed!

The quantum system is the global presence of *quantum impulses, quantum* of infinite low intensity.
Elementary processes create their own space-time by the speed value of quantum communication. The intensity corresponds to the quantum speed value.
Quantum systems with equal *quantum impulses* of infinite low intensity equally transfer *blue shift* impact of any quantum speed and intensity.

S. 1.7

1.7 The relation of the *quantum impulse* and the size of *space*

The quantum impulse is the one establishing space.

Ref. 1I3 With reference to 1I3

1L1

$$\lim q = \lim \Delta_{quantum} = (1-\zeta)\frac{dmc^2}{dt_i\varepsilon_e}\left(1-\sqrt{1-\frac{(c-i)^2}{c^2}}\right) = 0; \qquad \text{where } \lim_{0\to 1}\zeta = 1$$

1L2 and $0 < q < e_x$,

where e_x is the intensity of any of the electron processes, the stage of turning into neutron

1L3 process to be driven: $\lim e_x = 0$.

The question is: has the geometry of the space any impact on the intensities of the quantum impulse inside?
The answer is easy and simple: Elementary processes from the *plasma* to the *Hydrogen* process contain each other: the space-times of higher intensity contain the others with less intensities. The infinite large space-time of the plasma with its infinite high quantum speed and intensity contains all other space-times.

All events happen in space-times and all events generate quantum impulse. While space-times in relative terms contain each other, there is no way making separations in space. All elementary processes, tools or instruments, which could contribute to this kind of separation, are space-times themselves, with constant generation of quantum impulse.
Therefore there is no way having empty space. The accumulation of quantum impulses corresponds to the elementary processes. In the meaning of the generation of quantum impulse there is no difference in solid, liquid and gaseous statuses. Aggregate statuses of all kinds have their own space-time with their own quantum impulse generation and accumulation. There are no boundaries. The quantum space is a system, common for all elementary processes in any format of their aggregate status.
There are differences in space-time parameters, time counts, quantum speed and intensity values, but the energy of the quantum impulses they generate is equally $\lim q_{imp} = 0$.

2
Inflexion

We have to clarify the relation of the anti-proton collapse and the proton expansion at the anti-proton/proton inflexion. The intensities of the drives of the anti-proton collapses are equal, but they are acting with the variety of the durations. The intensities of the proton processes, following the inflexion this way are clearly different.

$$IQ_x = \frac{dmc_x^2}{dt_i \varepsilon_{x-}} = const\,; \qquad IQ_x = \frac{dmc_x^2}{dt_i \varepsilon_x} \neq const\,;\ \text{and} \qquad \text{2A1}$$

$$dt_{ix-} = dt_i \cdot \varepsilon_{x-} \qquad e_{px} = \frac{dmc_x^2}{dt_{ox}}\left(1 - \sqrt{1 - \frac{v^2}{c_x^2}}\right); \qquad \text{2A2}$$

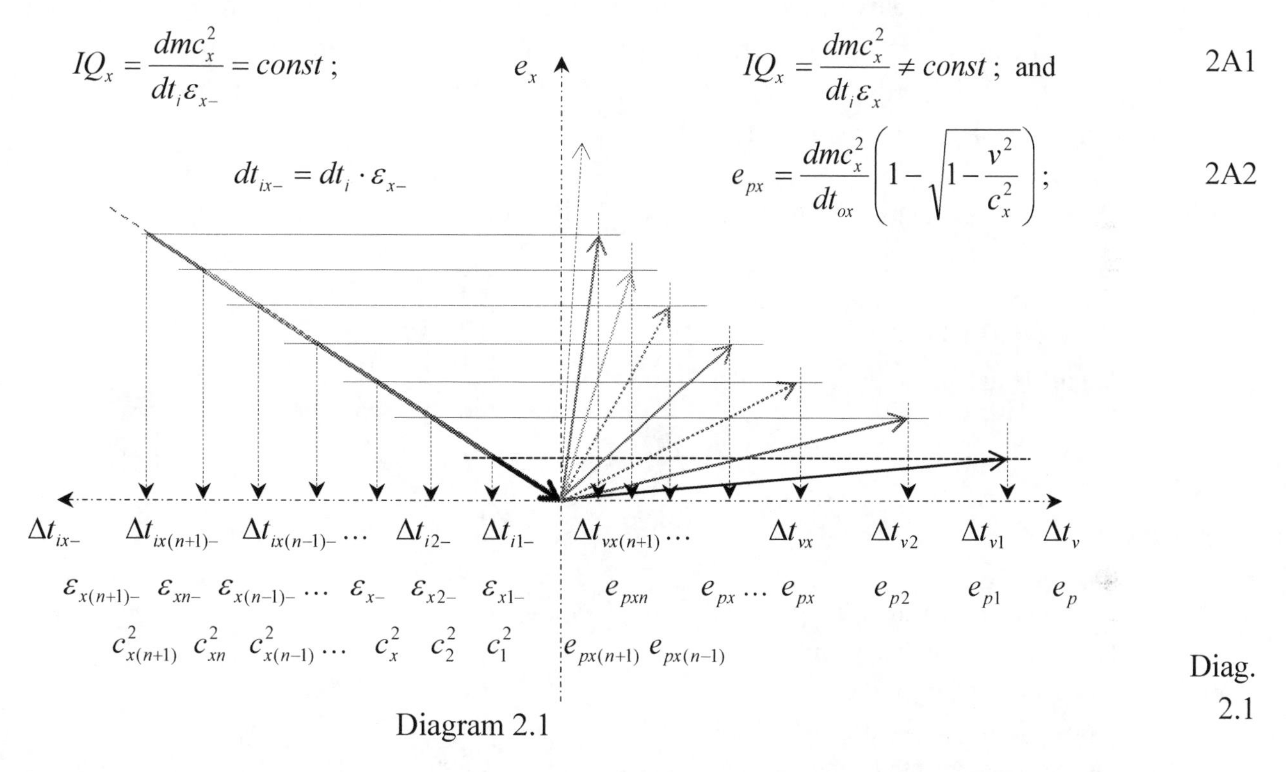

Diagram 2.1

Diag. 2.1

Diagram 2.1 above demonstrates the relation of the intensities of the incoming anti-electron processes (left hand side) and of the proton processes, leaving the anti-proton/proton inflexion (right hand side).

The intensities of the drives of the incoming collapses are equal for all elementary processes. Therefore the gradients of the arrows on the left hand side, representing the approach towards the *inflexion*, are congruent. The only difference is the duration of the processes.

Ref. 1B2- -1B5

With reference to 1B2-1B5, the higher the value of the quantum speed is, the higher is the number of the impacted quantum. The higher the quantum speed and the intensity coefficient of the anti-drive are, the longer is the duration of the drive and of the anti-proton process. [With the increase of the quantum speed, ε_{x-} the intensity coefficient of the anti-electron process is also increasing.]

The right hand side of the diagram is about the intensities of the proton processes. This is an approximation, since the proton process itself is a kind of parabola, with the reduction of the intensity, and with the constancy of the quantum speed values. Diagram 2.1 however demonstrates: the higher the value of the quantum speed is, the higher is the intensity (the gradient) of the proton process expansion at the start after the inflexion.

Ref. 1A2 1D2

With reference to 1A2 and 1D2, the intensity of the anti-electron processes (meaning: the number of the *quantum*, approached for the unit period of time), driving the anti-proton process towards the inflexion is equal for all elementary processes. These equal quantum drive intensities act however for different time counts, because the quantum speed values are different. The collapse of the anti-proton process is a kind of pumping up of the frequency: the intensity of the collapse is increasing, all the way on, reaching the inflexion. The number of the incorporating quantum impacts of the anti-electron processes is increasing. The intensity of the expansion on the other side starts with this accumulated during the collapse increased intensity.

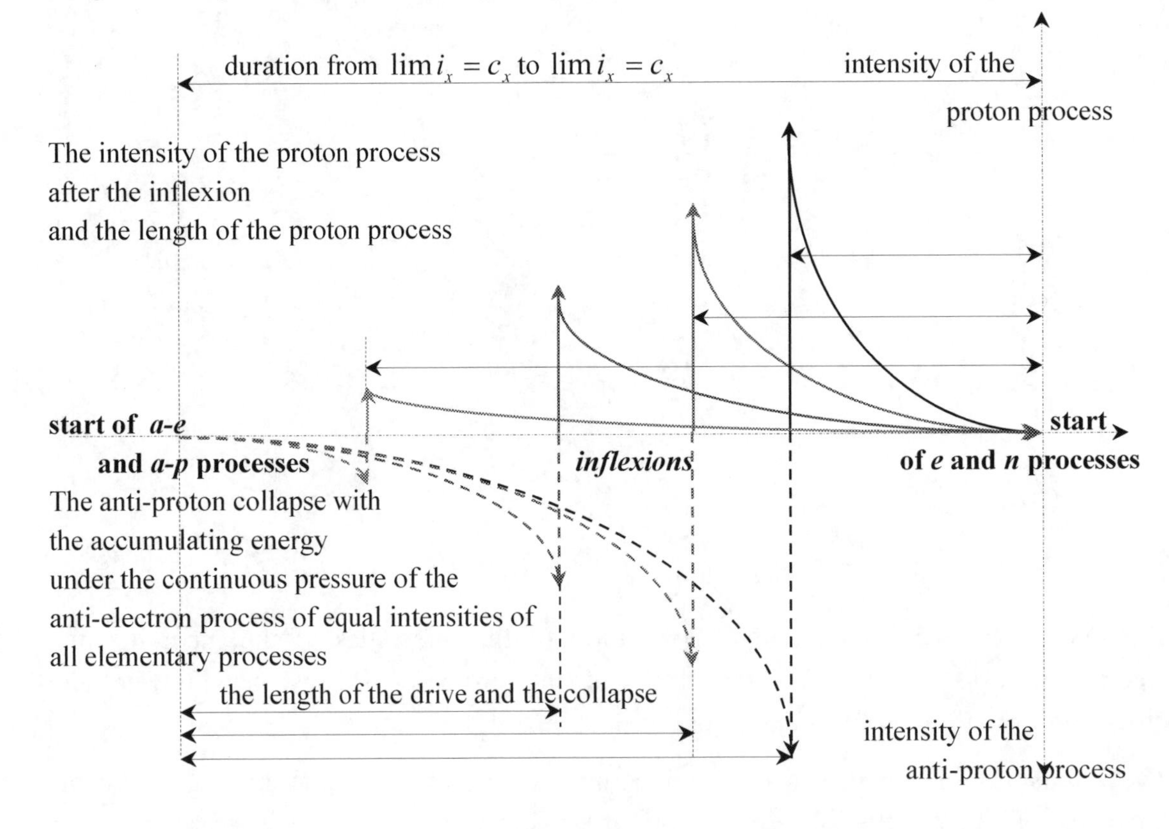

Diag. 2.2

Diagram 2.2

Ref. 1B11

The intensity of the anti-proton/proton inflexion corresponds to the quantum speed value, reference to 1B11, result of the generation of the quantum impulse. The energy transfer during the inflexion is without loss. Higher intensities need longer "pressing impacts of the drives" until the intensity of the anti-proton process finally is stepping over the inflexion. The intensities (frequencies) of the expansion of the proton processes are clearly different. The longer duration of the anti-electron process drive (and the anti-proton collapse) means, the number of the incorporating quantum impacts of the drive is more.

Ref. 1A2-1A6

With reference to Diagram 2.2, the proton process is repeating the intensity of the collapse. All proton processes reach $\lim i_x = c_x$, the speed of the expanded status, which belongs to equal time count! (Independently on the acting quantum speed values!) Ref. Diag. 1.1

The proton process:

$$\frac{dmc_x^2}{dt_p}\left(1-\sqrt{1-\frac{v^2}{c_x^2}}\right);$$

The higher the quantum speed is, the shorter is the time count approaching equal number of quantum during the process. 2A3

The "distance", the total number of quantum, approached by the expansion of the proton process between the *inflexion* and the start of the *electron process* are different,

- the number of the impacted quantum at the start are different, function of the quantum speed value: $e_{pxs} = \frac{dmc_x^2}{dt_{ox}}$; 2A4

and

- the *frequency* of the impacts at the end are also different, since while the time counts at $\lim i_x = c_x$ are identical, dt_i, different quantum speed values means different intensities : $e_{pxe} = \frac{dmc_x^2}{dt_i}$; 2A5
  (as the difference in the intensities of the quantum drives proves it)
- equal v speed values of the expansion belong to different time counts. The higher the value of the quantum speed is, the shorter is the time count. $dt_{pv} = \frac{dt_{po}}{\sqrt{1-\left(v^2/c_x^2\right)}}$; 2A6

Proton processes are similar, but the numbers of the impacted quantum during the processes are different! The time system of the electron processes are the same for all elementary processes, but the acting intensities are different, as $IQ_x \neq const$.

$$IQ_{x-} = \frac{dmc_x^2}{dt_i \varepsilon_{x-}} = const$$ 2A7

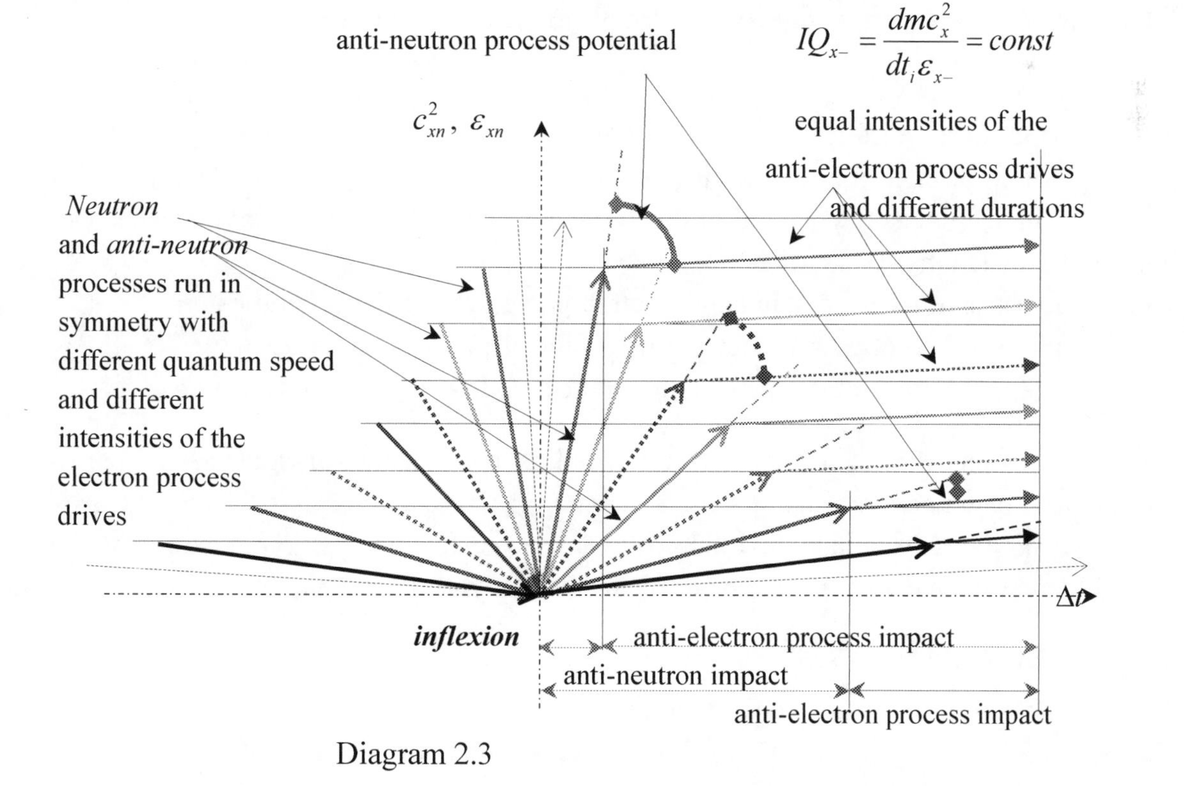

Diagram 2.3

Diag. 2.3

The neutron/anti-neutron *inflexion* is the classical example of the turn-around impact, as Diagram 2.3 on the previous page demonstrates it.

The symmetry of the process lines in Diagram 2.3 demonstrates the equal intensities before and after the inflexion and the continuity of the anti-neutron expansion. The anti-neutron process potential contains the intensity cover (transfer) of the anti-proton process and the generating anti-electron process surplus.
Diagram 2.3 is also presenting the difference in the duration of the anti-electron processes: The higher the intensity of the anti-neutron process is, the shorter is the anti-neutron process itself and longer is the duration of the anti-electron process.

Ref. 1B8

The intensity of the neutron/anti-neutron inflexion, relative to the anti-proton/proton inflexion is less and the reason is the automatic modification of the quantum speed value.

2A8

$$e_{nx} = \frac{dmc_x^2}{dt_i}\sqrt{1-\frac{(c_x-i_x)^2}{c_x^2}}\left(1-\frac{1}{\sqrt{1-\left(v^2/c_x^2\right)}}\right) = \frac{dmc_x^2}{dt_{ox}}\sqrt{1-\frac{(c_x-i_x)^2}{c_x^2}}\left(\sqrt{1-\frac{i^2}{c_x^2}}-1\right);$$

The equal intensities of the anti-electron processes result in energy transfer between the anti-neutron and the anti-proton processes of different durations. The energy transfer is regulated by the demand of the anti-proton process. The surplus of the cover potential of the anti-neutron process is released as the *quantum impact of gravitation.*

2B1

The intensity of the anti-proton process corresponds to the intensity of the anti-electron process *IQ* drive of $\frac{c_x^2}{\varepsilon_{x-}} = const$

2B2

The intensity of the energy source corresponds to the *IQ* drive of the acting electron process $\frac{c_x^2}{\varepsilon_x} \neq const$

This demand/source difference means constant surplus in the generating anti-electron processes of equal intensities (= equal to the quantum impact of gravitation) and results in the accumulating intensity potential of the anti-neutron process (= equal to the mechanical impact of gravitation).

2B3

The intensities of the developing anti-neutron processes and the generating anti-electron processes in the majority of the elementary processes are higher than the demand of the anti-proton/proton side, as: $\varepsilon_x = \frac{dt_n}{dt_p}$

This is the reason of the elementary evolution and the high number of elementary cycles of the elementary processes with increased quantum speed and increased electron process intensity values. Elementary evolution means the step by step reduction of the intensity/energy capacity of the elementary processes (cycles).

S. 2.0.1

2.0.1 Comment to the inflexions

The *two* top-charm-up (t-c-u) lines in the proton process results in two electron processes, *one* of which drives the down-strange-bottom (d-s-b) line, within the proton process. The *other* drives the d-s-b line within the neutron process.

One of the *two* d-s-b lines within the neutron process is driven by the electron process, result of the *single* t-c-u line within the neutron process. The *other* d-s-b line is driven from the t-c-u of the proton process.
Inflexions are global events, which happen in three parallel lines in each elementary process. The anti-electron processes of the b-s-d expansions of the anti-neutron process drive its own u-c-t collapse and the u-c-t collapse of the anti-proton process. The anti-electron process of the anti-proton b-s-d drives its own u-c-t into collapse.

It is better to speak about u-c-t/t-c-u and d-s-b/b-s-d inflexions than anti-proton/proton and neutron/anti-neutron inflexions in the elementary processes. These inflexions within the elementary process are similar, as they are similar steps of the elementary cycles and run in parallel, just happen in different composition of the cycles.
The number of the elementary cycles running in parallel is infinite large, with estimated duration of the cycles of 10^{27-29} years within our space-time.

2.1 The *proton processes* are similar, with the appearance of this similarity within different space-times with different quantum speed and electron process intensity values

S. 2.1

The intensity of the electron process at the direct side corresponds to the intensity of the *quantum membrane*, generated by the anti-electron process surplus at the anti-side.
The higher the pressure of the quantum membrane of the surplus in the anti-process side is, the higher is the intensity on the direct side.
The reason of the surplus is the increased intensity of the anti-neutron process, the increased generation of the anti-electron processes.

$\frac{\varepsilon_{x-}}{\varepsilon_x} = n_x$; at the same time $\varepsilon_x = \frac{1}{\varepsilon_{x-}}$; and this way $\varepsilon_x = \frac{\varepsilon_{x-}}{n_x}$; and $n_x = \varepsilon_{x-}^2 = \frac{1}{\varepsilon_x^2}$; 2C1

Making the substitution, we are coming to the balance of the electron and the anti-electron processes on the direct and the indirect sides:

$$\frac{dmc_x^2}{dt_i\varepsilon_x}\left(1-\sqrt{1-\frac{(c_x-i_x)}{c_x^2}}\right) = n_x\frac{dmc_x^2}{dt_i\varepsilon_{x-}}\left(1-\sqrt{1-\frac{(c_x-i_x)}{c_x^2}}\right); \text{ and } c_x^2\cdot\varepsilon_{x-} = \frac{c_x^2}{\varepsilon_x} \qquad \text{2C2}$$

The anti-process is impacting the direct process and establishing the elementary balance.

If the intensity of the anti-electron process is: $\varepsilon_{x-} > 1$, 2C3

the intensity of the generation of the surplus on the anti-side is: $\varepsilon_{ex-} = \varepsilon_{x-} - 1$; 2C4

In the case of $\varepsilon_{x-} = 1$, there is no surplus.

If the generation of the number of the anti-electron processes is n_x, the number of the anti-proton processes driven, meaning the number of the next elementary cycles, with reference to 2C4 is: $n_{n+1} = n_n\left(1-\frac{\varepsilon_{x-}-1}{\varepsilon_{x-}}\right)$; 2C5

2C5 represents neutron process dominant elementary processes.

2C6 With identical logic, just for proton process dominance, with reference to 2C5 is: $n_{n+1} = n_n \left(1 - \frac{\varepsilon_x - 1}{\varepsilon_x}\right)$;

2C5 and 2C6 above are not about losing on elementary cycles.

Ref. Table 3.1 2C4 represents the intensity of the reduction of the elementary cycles. The intensity of the elementary cycle is establishing the gradient of the reduction of the internal cycles. With reference to Table 3.1, for example the reduction of the elementary cycles in the *Oxygen* process is very slow. The reduction within the *Uranium* process is at much-much higher rate. The reduction within the *plasma* is of infinite high gradient.

2C5 (for neutron dominance) and 2C6 (for proton dominance) also mean the intensity of the rebirth of the elementary cycles. In the case of the *plasma* it is infinite high, in the case of the *Hydrogen* process it is of infinite low rate.

In the case of neutron process dominant elementary processes, the internal conflict of the generating surplus of the anti-electron processes (the quantum membrane) establishes the intensity of the electron process and generates the quantum impacts of gravitation.

In the case of proton process dominance, the surplus is generating on the direct side and initiates the quantum communication of the elementary processes.

The completion of the elementary cycle means the release of the surplus of the anti-electron or/and the electron processes. This is the point when the intensity difference between the proton and the neutron processes is being resolved and the balance of the new cycle becomes established.

Ref. 1B11 The anti-electron process controls the elementary process, reference to 1B11, with slight reduction of the speed value of quantum communication.

The anti-proton/proton inflexion establishes the space-time of the elementary process.

The quantum speed of the elementary cycles becomes of less and less value.

The proton processes as events, for all elementary processes are similar, with the variety of the quantum speed and the intensities values. Having the variety of the speed of quantum communication, elementary processes have their own space-times.

2D1 The proton process as event is represented by the expanding acceleration: $v = a \cdot \Delta t$

The proton process of elementary process x is:

2D2 $e_{px} = \frac{dmc_x^2}{dt_{px}} \left(1 - \sqrt{1 - \frac{v^2}{c_x^2}}\right)$; with the actual time count of the event: $dt_{act} = \frac{dt_{px}}{\sqrt{1 - \frac{v^2}{c_x^2}}}$;

Ref. 1A3

2D3 With reference to 1A3, the time count of the quasi rest status, closest to the inflexion (dt_o) is equal for all elementary processes: $\lim \Delta t_o = 0$

dn_x means the full number of the impacted quantum, as the proton process starts in its full expanding intensity at the *inflexion*, with the step by step decrease of its intensity.

With reference to 2D1, the changing $v = a \cdot \Delta t$ speed value of the acceleration is the one representing the actual time count of the proton process.

The intensity demand of the collapsing quark processes on the anti-proton sides are equal, the drives are $IQ_{x-} = const$. The quark processes of the expansion and the collapse are equal in elementary processes, just of different quantum speed and intensity values.
For neutron process dominant elementary processes the balance of the direct side is in order:

$$\frac{dmc_x^2}{dt_p\varepsilon_p}\left(1-\sqrt{1-\frac{i_x^2}{c_x^2}}\right)=\zeta\frac{dmc_x^2}{dt_n\varepsilon_n}\sqrt{1-\frac{(c_x-i_x)^2}{c_x^2}}\left(1-\sqrt{1-\frac{i_x^2}{c_x^2}}\right); \qquad 2E1$$

But for the anti-direction:

$$\frac{dmc_x^2}{dt_{n-}\varepsilon_{n-}}\sqrt{1-\frac{(c_x-i_x)^2}{c_x^2}}\left(1-\sqrt{1-\frac{i_x^2}{c_x^2}}\right)>\frac{dmc_x^2}{dt_{p-}\varepsilon_{p-}}\left(1-\frac{(c_x-i_x)^2}{c_x^2}\right)\left(1-\sqrt{1-\frac{i_x^2}{c_x^2}}\right); \qquad 2E2$$

Elementary processes with higher quantum speed values have larger space-times and less relative values of acting anti-electron process intensities. High quantum speed space-times with increased c_x quantum speed in the counter and with increased intensity coefficient ε_{x-} in the denominator mean limited number of impacts and as consequence, the gradient of the generation of the elementary cycles is relatively low value. The surplus is for the quantum impact of gravitation!

For keeping the elementary process stable, the number of the operating parallel elementary cycles should be of high number.

For proton process dominant elementary processes, the absolute balance is guaranteed in the anti-directions. The proton process is generating more energy/intensity value than it is utilized by the neutron process. The surplus is spent for quantum communication.

$$\frac{dmc_x^2}{dt_p\varepsilon_p}\left(1-\sqrt{1-\frac{i_x^2}{c_x^2}}\right)>\frac{dmc_x^2}{dt_n\varepsilon_n}\sqrt{1-\frac{(c_x-i_x)^2}{c_x^2}}\left(1-\sqrt{1-\frac{i_x^2}{c_x^2}}\right); \qquad 2E3$$

For the anti-direction:

$$\frac{dmc_x^2}{dt_{n-}\varepsilon_{n-}}\sqrt{1-\frac{(c_x-i_x)^2}{c_x^2}}\left(1-\sqrt{1-\frac{i_x^2}{c_x^2}}\right)=\xi\frac{dmc_x^2}{dt_{p-}\varepsilon_{p-}}\left(1-\frac{(c_x-i_x)^2}{c_x^2}\right)\left(1-\sqrt{1-\frac{i_x^2}{c_x^2}}\right); \qquad 2E4$$

The value of the surplus at the direct side (of the 8 proton process dominant elementary processes) is very limited therefor there is no need for high number of elementary cycles.

2.2 The quantum impact of the motion

S. 2.2

The *EVENT* in "quantum case" means the number of the impacted quantum at the inflexion, with certain quantum speed value, initiated by the collapse on the anti-side.
If elementary processes (including the inflexion itself) are in motion, with constant speed or acceleration, the intensity of the process will be influenced, depending on the drive of the motion. Whether it is internal or external.

There are here two basic differences in the time counts, caused by motion:

1. The event is taken by the motion – by *external* drive	2. The event itself is also the drive of the motion – *internal* drive:
2F1 $w_{ext} = \frac{EVENT}{dt_x\sqrt{1-\frac{v^2}{c^2}}} - \frac{EVENT}{dt_x};$	2F2 $w_{\text{int}} = \frac{EVENT}{dt_x} - \frac{EVENT}{dt_x}\sqrt{1-\frac{v^2}{c^2}}$
2F3 As $dt_{xv} = dt_x\sqrt{1-\frac{v^2}{c^2}}$ — The time count is slowing down as the event is driven for the count of another energy source	2F4 As $dt_{xv} = \frac{dt_x}{\sqrt{1-\frac{v^2}{c^2}}}$ — The time count is speeding up, as the event covers the drive from its energy reserve as well

In the case of identical events in motion, the motion is influencing the time count of the events.

2F5 The *EVENT* has been characterised by the intensity of the electron process: $\frac{dmc_x^2}{dt_i\varepsilon_x}\left(1-\sqrt{1-\frac{(c_x-i_x)^2}{c_x^2}}\right);$

In the case of *external* drive the time count difference is valid only for the period of the acceleration. (While the acting impact of the acceleration clearly defines the system in motion.) There is no way at constant speed v of the event to decide which system is in relative motion, relative to the other. The time count relation of the two systems is reciprocal. It means in this case, they are of the equal status. (Ref. 2A5)	The *internal* drive specifies the system in motion. The time count becomes *quicker* – as the intensity/energy of the event is *decreasing*, leading to the *decrease* of the intensity of the communication with quantum impulses! The event itself is still about n_x quantum impact, but as the drive also needs energy, the IQ_x capacity value becomes reduced, the time count is increasing.

The key is the electron process, the work process of the *EVENT*.

A./ If the motion (acceleration) is driven by the internal energy of the process:

2F6
$$w_e = \frac{dmc_x^2}{dt_i\varepsilon_x}\left(1-\sqrt{1-\frac{(c_x-i_x)^2}{c_x^2}}\right)\left(1-\sqrt{1-\frac{v^2}{c_x^2}}\right) =$$

2F7
$$= \frac{dmc_x^2}{dt_i\varepsilon_x}\left(1-\sqrt{1-\frac{(c_x-i_x)^2}{c_x^2}} - \sqrt{1-\frac{v^2}{c_x^2}} + \sqrt{1-\frac{(c_x-i_x)^2}{c_x^2}}\sqrt{1-\frac{v^2}{c_x^2}}\right);$$

these are the components, spent for the drive and for the motion	this is the portion remaining for the collapse

With reference to the above the neutron process will be:

2F8
$$\xi\frac{dmc_x^2}{dt_n}\sqrt{1-\frac{v^2}{c_x^2}}\sqrt{1-\frac{(c_x-i_x)^2}{c_x^2}}\left(1-\sqrt{1-\frac{i_x^2}{c_x^2}}\right);$$

with the actual cover demand from the proton process of: $e_n = \frac{dmc_x^2}{dt_n}\sqrt{1-\frac{v^2}{c_x^2}}$; 2F9

With reference to the anti-process, the intensity of the collapse at the anti-proton/proton inflexion is:

$$e_{p-} = e_p = \frac{dmc_x^2}{dt_p}\left(1-\frac{(c_x-i_x)^2}{c_x^2}\right)\sqrt{1-\frac{v^2}{c_x^2}}; \quad \text{2G1}$$

and the quantum speed of the new cycle (elementary process) is: $c_{x+1} = c_x\left(1-\frac{(c_x-i_x)^2}{c_x^2}\right)\sqrt{1-\frac{(a\cdot\Delta t)^2}{c_x^2}}$; 2G2

With permanent $a = const$ acceleration, each cycle reduces the intensity of the original $e_p = \frac{dmc_x^2}{dt_p}$; 2G3

The decrease of the quantum speed is: $c_{x(n+1)} = c_{xn}\left(1-\frac{(c_{xn}-i_{xn})^2}{c_{xn}^2}\right)^n \sqrt[2n]{1-\frac{(a\cdot\Delta t)^2}{c_{xn}^2}}$; 2G4

B./ In the case the source of the motion (acceleration) is external – following the logic of the process above, the intensity of the electron process becomes increased:

this means the increase of the energy potential of the electron process: $e_e = \frac{dmc_x^2}{dt_i\varepsilon_x}\left(\frac{1}{\sqrt{1-\frac{v^2}{c_x^2}}}-1\right)\left(1-\sqrt{1-\frac{(c_x-i_x)^2}{c_x^2}}\right)$; 2G5

2G5 means, the electron process starts from and increased value: $IQ_x = \frac{c_x^2}{\varepsilon_x\sqrt{1-(v^2/c_x^2)}}$; 2G6

The electron process runs out to the generation of the quantum impulse with $\lim q = 0$, just the quantum speed value as result of the external acceleration has become an increased value. This way the intensity potential of the electron process becomes increased.

The energy of the anti-proton/proton inflexion is: $e_{p-} = e_p = \frac{dmc_x^2}{dt_p\sqrt{1-\frac{v^2}{c_x^2}}}\left(1-\frac{(c_x-i_x)^2}{c_x^2}\right)$; 2G7

The benefit of the external acceleration is the increase of the operating quantum speed and the loss on this increased value : $c_{x(n+1)} = \frac{c_{xn}}{\sqrt[2n]{1-\frac{(a\cdot\Delta t)^2}{c_{xn}^2}}}\left(1-\frac{(c_{xn}-i_{xn})^2}{c_{xn}^2}\right)^n$; 2G8

The anti-proton/proton and the neutron/anti-neutron inflexions are consequences. The transformation of the intensities follows the electron/anti-electron process drives.
The number of the impacted quantum at the anti-proton/proton inflexion is determined by the intensity of the electron process.

C./ <u>The best example</u> for the *external* drive is the speeding up of the *Hydrogen* process. As result, the *Hydrogen* process remains *Hydrogen* process. But the acceleration happens within the space-time of the *Earth* and increases the intensity of the *Hydrogen* process.

2H1 $$\frac{dmc_x^2}{dt_i \varepsilon_x \sqrt{1-\frac{v^2}{c_E^2}}}\left(1-\sqrt{1-\frac{(c_x-i_x)^2}{c_x^2}}\right);$$ The acceleration increases the quantum impact of the *Hydrogen* process.

(The speed of the quantum communication of the *Hydrogen* process has already been increased by the quantum impact of *gravitation* to the quantum speed of the *Earth* surface.)
The increase of the quantum speed means the increased intensity of the quantum drive.
The increase of the *IQ* drive in 2H1 keeps the elementary process without change, as the increase of the *IQ* does not impact the intensity relation of the proton and the neutron processes, the sphere symmetrical accelerating collapse and expansion.

S. 2.3

2.3
Parallel events with different time counts

The appearance of events means the quantum impact they do.
The appearance of identical events in different space-times is different.

Events with increased intensity do not just happen in their own space-times of increased intensity for shorter time count, but have the intensity capacity to appear in space-times of less intensities and may cause there conflicts.
The impacts of events with moderate intensity are less "aggressive". These events happen in their own space-times for longer time count.

2I1

$\Delta t_1 = \frac{n_x}{IQ_{x1}}$; and $\Delta t_2 = \frac{n_x}{IQ_{x2}}$; $IQ_{x1} > IQ_{x2}$.	The numbers of the impacted n_x quantum impulses in both cases are equal; but the quantum drives are different. This way: $\Delta t_1 < \Delta t_2$

While in space-time (1) the event happens for less,
the same event in space-time (2) in parallel happens for longer time count.
Space-time (1) is of more intensity than space-time (2).

The number of the impacted quantum decides what the time flow of the events is. Events with the *same number of the impacted quantum* in principle are the same, but they may happen for different time counts in space-times of different quantum speed values.

All we have our own space-time, with our own specific intensities and quantum speed values. The external quantum space above the *Earth* surface has its own time-count as well. There are natural events identical and common for all of us, like the rotation of the *Earth* and orbiting around the *Sun*. They are the practical tools for comparing events and counting time, passed on the *Earth* surface. But this is not our personal internal time count!

Our personal time count follows our internal elementary structure, the intensity of our personal integrated elementary operation.

Our personal *space-time* generates our personal (dt_x) time count.

$$\frac{dn_x}{dt_x}q = \frac{dmc_x^2}{dt_i\varepsilon_x}\left(1-\sqrt{1-\frac{(c_x-i_x)^2}{c_x^2}}\right); \qquad \frac{dn_x}{dt_x}q = \frac{dm}{dt_i}IQ_x\left(1-\sqrt{1-\frac{(c_x-i_x)^2}{c_x^2}}\right); \qquad 2I2$$

the intensity of the impacted quantum — our personal *blue shift* impact

Our personal quantum communication is based on our personal, but aggregate c_x quantum speed and ε_x personal intensity. We are impacting n_x quantum impulse within our personal *space-time* by our personal elementary processes. The personal space-time of human living systems is based on infinite high number of elementary processes.
While the quantum speed and the intensity values of these elementary processes vary, they are all within a certain range of the time systems of the elementary processes.
Elementary processes communicate and each of the living systems has an aggregate impact on the external quantum system. The same way, as minerals do.

$$\lim q = \lim \frac{dmc^2}{dt_i\varepsilon}\left(1-\sqrt{1-\frac{(c-i)^2}{c^2}}\right) = 0 \qquad 2I3$$

c – is the *variety of quantum speed values*
ε – is the *variety of intensities.*

$$\frac{dn_x}{dt_x}\frac{dmc^2}{dt_i\varepsilon}\left(1-\sqrt{1-\frac{(c-i)^2}{c^2}}\right) = \frac{dmc_x^2}{dt_i\varepsilon_x}\left(1-\sqrt{1-\frac{(c_x-i_x)^2}{c_x^2}}\right); \longrightarrow \frac{dn_x}{dt_x}\frac{c^2}{\varepsilon} \equiv \frac{c_x^2}{\varepsilon_x} \text{ and} \qquad 2I4$$

from 2I4 follows: $dt_x = \frac{dn_x}{IQ_x}$ ⟵ n_x is the number of the impacted quantum; ⟵ is the IQ of our personal quantum "drive".

$$\Delta t_x = \frac{n_x}{|IQ_x|} = \varepsilon_x\frac{n_x}{c_x^2}; \qquad 2I5$$

Time means the number of the impacted quantum, divided by the quantum drive.

Time represents the intensity of the quantum communication with the quantum system!

- The number of the impacted quantum for the unit period of time means our personal space-time.
- At equal number of impacted quantum, the higher the value of the *IQ* drive is, the less is our time count.
- At equal time counts, the higher the quantum drive is, the higher is our quantum impact.

The higher the intensity of our quantum communication (IQ_x) is, the less (slower) is our personal time count!

Our quantum impact should be understood as a tendency, about a greater or lesser space-time. The *blue shift* impact within the brackets in 2I2 and equally in 2I3 are all of infinite low values themselves. Therefore it is better only to relate to the number of the impacted quantum.

Ref. 2I2 2I3

The slowdown of the time count is only possible if the IQ_x is of increasing value. =

= The increase of the IQ_x value results in the slowdown of the time count.

Quantum communication with quantum systems is natural need. Natural need for the elementary processes, for events of any kind and for living systems (including human)!

This natural need of quantum communication is based on two principles:
- on the *feed* of the quantum system – by the constant generation of quantum impulse; and
- on the *load* of the quantum membrane – by the *blue shift* impact.

The quantum system is the aggregate sum total of the quantum impulses, energy quantum. There is no space-time (quantum system) theoretically without load, as the quantum impact of the plasma is of infinite high intensity; the space-time of the plasma contains infinite large number of quantum impulses; the space-time of the plasma is infinite large. Therefore the generating anti-electron process *blue shift* surplus of the plasma is impacting infinite high number of quantum impulses. *The space-time of the plasma contains all other space-times of the elementary evolution with less quantum speed values.*

The point here is that the quantum impacts of the elementary processes of the elementary evolution are in conflict with the quantum impacts of the plasma, slowing down the quantum speed of the quantum impact of the plasma and creating certain material composition, discussed later on.

This means *plasma* and the processes of the elementary evolution generate space-times, but the *load* of the space-times is limited and might be missing.

Each cycle of elementary processes *feeds* the quantum system.
The *load* of the quantum membrane means the "tightness" (the number) of the quantum impacts acting in parallel within the quantum system.
In normal natural circumstances (without any IT tools) the *load* is coming
- from the quantum impact of *gravitation*, the anti-electron process *blue shift* impact of elementary processes, and
- from the eight elementary processes (*H, He, Ni, O, C, Si, Ca, S*).
  The electron processes *blue shift* impact of these 8 elementary processes increases the load, established by the quantum impacts of gravitation.

With reference to 2I5 and the number of the impacted quantum impulses:

2I6 $$\Delta t = \frac{n_x}{IQ_x}\text{; means } f = \frac{1}{\Delta t} = \frac{IQ_x}{n_x};$$ With the increase of the quantum impact the frequency of the quantum membrane is increasing.

As the increase of the space means the increase of the numbers of the quantum impulses, the intensity of the load of the system, with the increase of the highs above the *Earth* surface is decreasing. Here again in 2I6 the tendency of the change is the one to take into account. (The number of quantum impulses is infinite high anyway and the calculated intensities give always of infinite low values.)
The aura of living systems is a kind of personal quantum membrane, surrounding the living system, having its clear boundaries. The quantum impact prevails inside, conflicting with the external quantum impact at the boundaries. If the intensity of the personal impact is

higher, the conflict (the aura) is shining. It is bright. If the intensity of the external impact is the one, which is higher, there is no conflict and no sign of brightness.

Living systems develop their own space-times, based on the integrated impacts of their elementary processes. As the quantum speed value is the distinguishing tool of the space-times, space-times have their own speed of quantum communication. Each IQ_x value is establishing an elementary *inflexion* of certain quantum speed and intensity values. *Inflexions* with equal IQ_x drives and equal quantum speed values mean not just similar elementary processes, but similar space-times as well.

This ensures that quantum impacts or signals of space-times of equal quantum speed and intensity values may easily find each other; they may have contact with each other and may communicate at the common platform of quantum speed and intensity values. (As with reference to Section 4.3.1, the small scale experiments with pyramids prove it in practice.) This still needs however intensity/energy capacities at the necessary level in order to overcome all *blue shift* conflicts on the way, finding the partner space-time.

Ref. S. 4.3.1

S. 2.3.1

2.3.1 In addition to the subject

The source of the sphere symmetrical expanding acceleration of the *Earth* surface is the *plasma* process with its infinite high quantum speed and intensity. These two parameters, with the constant utilisation of the internal intensity reserves of the *Earth* are decreasing. This is the reason of the elementary evolution and the formation of the *Earth* core with minerals.
We, human beings on the *Earth* surface have been in motion, corresponding to the actual speed value of the expansion of the surface. This speed value is the one determining our personal time count on the *Earth* surface.
The only way for slowing down our personal time count is to increase our IQ value.

As with reference to *The Quantum Impulse and the Space-Time Matrix*

Ref. QISM

$$c_x^2 \cdot \varepsilon_x = \frac{c_x^2}{\varepsilon_{x-}} = const$$

the intensity of the anti-electron process quantum drives for all elementary processes are of equal value. 2J1

The quantum impacts around us have their effect on the intensity of our quantum communication. The demand to withstand this high "intensity pressure" of the external quantum system creates conflict. This automatically increases our internal work load.

Our existence is constant quantum communication with the impacts of the quantum system around us. The *less* is the loss of our internal intensity (internal energy), the slower is our time system, consequently, the more is the intensity of our quantum communication.
Our internal work (the *EVENT*) and our motion from external source in parallel changes the time relations only if the external drive means constant acceleration:

$$w_{ext+\text{int}} = \frac{EVENT}{dt_x\sqrt{1-\frac{(a\Delta t)^2}{c^2}}}\sqrt{1-\frac{v^2}{c_x^2}} - \frac{EVENT}{dt_x};$$

In the case of $a\Delta t \geq v$ there is either no loss or the intensity of the process is increasing. 2J2

The less is the intensity demand of the external quantum system, the more efficient is our personal quantum communication.
With the increase of the quantum speed, the time count is slowing down. The slowest and the most intensive process is the *plasma*. The less intensive process, with the quicker time count is the *Hydrogen* process.

Ref. S.9

In the case of any external drive of elementary processes, the intensity is increasing. This is the principle, with reference to Section 9, of the energy generation by accelerating the *Hydrogen* process: The conflict of the electron process with the quantum impact of *gravitation* generates heat.

3
Elementary communication

The elementary processes are the appearances of the original plasma process, just in space-times of much-much less intensity and of less quantum speed values, close to our speed of quantum communication, the speed of light on the *Earth* surface. Are they in solid, gaseous or liquid aggregate status, makes no difference. Separately or in their communication they represent the process of the elementary evolution.

Keeping pieces of minerals in our hands, pouring liquids, blowing gases the case is the same. Infinite numbers of elementary processes are in progress. Elementary processes represent the cooling down status of the plasma.

Liquids and gases represent the natural conflict of the proton process dominance, specific step of the elementary evolution, or results of intensive external impact. Minerals in solid status can be melted, can be separated on the basis of the increase of the intensity of the electron process; generating this way conflict.

The point here, in this section and the next is to prove, all elementary communication we speak about is not just theory – it is in fact proven by practice and happens in front of us. The existing difference in space-times however makes it difficult for us to recognise it.

When the elementary processes of the Periodic Table represent an estimated lifetime of 10^{27-29} years in our space-time, it is very difficult to follow the case and realise the real process. Our space-time on the *Earth* surface has been determined by the quantum speed on the *Earth* surface, $c_{Earth} = 299792$ km/sec.

The sphere symmetrical expanding acceleration of the *Earth* surface is $\lim i_{Earth} = c_{Earth}$. It means *infinite long* time count:

$$dt_{i-Earth} = \frac{dt_o}{\sqrt{1-\frac{i_E^2}{c_E^2}}} = \infty ; \qquad \text{3A1}$$

And we cannot be surprised having in our practice the estimated lifetime of the elementary stages of this quasi infinite long category.

Elementary communication is energy/mass intensity exchange on the basis of
- the identical time system of the electron processes;
- the fact that neutron processes have been reciprocally driven by the electron processes of the communicating elementary processes.

Elementary processes have their own specific balanced statuses. Elementary evolution produces infinite large variety of elementary structures, but only those are identified as balanced, where the elementary cycles ensure the fluent continuity of the process: the repetition of the elementary cycles with decreasing quantum speed value.

We obviously cannot wait out in our life the duration of an elementary cycle. But there are infinite numbers of elementary cycles running in parallel. And there are always elementary cycles in different stages of the elementary progress to be found, including the inflexions, the start and the end of the cycle. Therefore we are in the position to measure the phases of elementary processes; we will be always capable to measure the different statuses between the start and the completion of elementary cycles.

The *plasma* and the *Hydrogen* processes are the two ends. Having the neutron process dominance of our known elementary processes in the Periodic Table, we are still closer to the *plasma* status in the elementary evolution. But the best period is, when the two ends with proton and neutron process dominance have the chance to communicate.

The objective of the elementary communication is the reciprocal use of the strengths of the elementary processes, reciprocal assistance in order to reach a status, as closer as possible to equilibrium. Balanced elementary status means an operation without electron process surplus or deficit.

S. 3.1

3.1 The relation of the *quantum impulse* and the intensity of the elementary process

We can identify a so called unit number of the operating elementary cycles. This unit number characterises the intensity of the quantum drive available within the elementary process.
The less the intensity of the electron process is, the less is the surplus of the anti-electron processes. The higher the unit number of the elementary cycles is, the higher is the number of the generating *quantum impulses, quantum*. (The *Hydrogen* process is an exemption, since the cycle in this elementary process is not completed.)
The higher the number of the elementary cycles is, the higher is the anti-electron process surplus, the higher is the number of the generation of the *quantum impact of gravitation* (on the indirect side of the elementary process).
The increase of the number of the acting electron processes is equivalent to the increase of the intensity of the *IQ* drive. Increased intensity means conflict.

3A2 $$n\frac{dmc_x^2}{dt_i\varepsilon_x} = \frac{dmc_x^2}{dt_i\varepsilon_x\sqrt{1-\frac{v^2}{c_x^2}}}$$

The conflict is equivalent to the speeding up of the elementary process. The acceleration and also the increase in numbers are generating conflict, changing the *IQ* drive value. Constant acceleration means permanent conflict.

Elementary communication is initiated by conflicts. The number of the quantum drives of the elementary process without conflict, in normal circumstances is equal to the number of the elementary cycles. This corresponds to $n = 1$ in equation 3A1 above.

3A3 In the case of conflict the meaning of n_{unit} is a relation of: $n_{unit} = \frac{n_{conflict}}{n}$

What does the unit number of the electron processes within a space-time mean?

1. the number of the cycles is equal to the number of the generating electron process quantum drives without surplus;
2. elementary processes with electron process surplus have space-times in conflict as natural characteristics of the elementary process;
3. the number of the cycles is equal to the number of the generated quantum impulses.

The quantum space or just space means the volume of the accumulated *quantum impulses, quantum, quanta* of infinite low intensity and of infinite long duration. The *space* is neutral. The *space* is expanding as the generation of the quantum impulses is permanent. Quantum impulses transfer all signals without change. The *space* is not generating any events – it contributes only to the operation of space-times, transfers all quantum impacts.

Space-time is "the space and the time system" of the elementary processes in line with their own intensity and quantum speed values. Elementary processes establish their own *space-times* within the *space*. *Space* is the integration of all existing space-times.
Space-times belong to elementary processes. Elementary processes within the *space* happen with different quantum speed values and intensities. The main characteristic of the space-time of the elementary processes is the *quantum speed*, (which includes in fact the dimensions of both: the space and time). Infinite high quantum speed means infinite large space-time with reaching infinite large number of quantum impulses for infinite short time. Infinite low quantum speed means infinite long time count. The intensity of the *space-time* can be impacted and changed.

One and the same event, happening in different *space-times* have different durations.
Observing an event happening within a certain space-time means the projection of the event on the space-time of the observation. The quantum speed and the time count of the space-time of the observation might be different than the time count of the space-time, where the event in fact happens.

Elementary processes with high quantum speed and increased intensity need extra high external quantum impact for generating conflict. For example, the boiling, as conflict, for elementary processes with increased intensity, like *Iron* needs increased external impact.

The quantum impact of gravitation is generated by the anti-electron processes surplus of the elementary evolution of the *Earth*.

means: number

#e the accumulating and the acting potential of the electron and the proton process surplus; ← proton process dominance

#(p) =#(e) = #(n) =#(n-) =#(e-) = #(p-)= #(p) 3A4

neutron process dominance → **#e-** the surplus of the anti-electron and the anti-neutron processes, the impact of *gravitation*;

External quantum impact might generate conflict with the quantum impact of gravitation.

3A5 $c_x^2 \cdot \varepsilon_x = \frac{c_x^2}{\varepsilon_{x-}}$ The *IQ* drives of the anti-proton processes are equal, in order to ensure the similar character of the proton processes.

The generating *blue shift* impact of the surplus of the anti-electron processes is the origin of the quantum impact of *gravitation*. The cover potential of the anti-neutron process of the surplus is the mechanical impact of *gravitation*, the intensity of the expansion itself.

Each elementary process contributes to the *quantum impact of gravitation*, function of the quantum speed and the intensity of the anti-electron process. The *quantum impact of gravitation* of elementary processes with increased quantum speed is not just more in numbers, but these impacts are acting within a space-time of increased intensity. Elementary processes with increased quantum speed have ($\varepsilon_{x-} >> 1$) increased coefficient of anti-electron process intensity, increased acting intensity and increased quantum impact of gravitation!

3A6
$$\frac{c_o^2}{\varepsilon_{o-}} = const = \frac{c_x^2}{\varepsilon_{x-}}; \Rightarrow \varepsilon_{x-} = \varepsilon_{o-}\frac{c_x^2}{c_o^2}; \Rightarrow e_{gr} = \frac{dmc_x^2}{dt_i\varepsilon_{x-}}\left(1-\sqrt{1-\frac{(c_x-i_x)^2}{c_x^2}}\right);$$

The impact of anti-electron processes is continuous; the intensity of the impact is permanent. While the intensities of the quantum drives of the anti-electron processes of all elementary processes are identical, the duration of the impact vary. The intensity cover of the anti-proton collapse is coming from the anti-neutron expansion, at the quantum speed of the elementary process. This is the impact resulting in the inflexion and the one establishing the intensity of the proton process.

The generation of *quantum impulses* is permanent; the value of the *quantum speed* is decreasing step by step. The quantum impact of gravitation of the plasma with infinite high quantum speed value is acting for infinite long time.

3A7
$$\frac{dmc_x^2}{dt_{on-}}\sqrt{1-\frac{(c_x-i_x)^2}{c_x^2}}\left(1-\sqrt{1-\frac{i_x^2}{c_x^2}}\right) \rightarrow \frac{dmc_x^2}{dt_{op-}}\left(1-\frac{(c_x-i_x)^2}{c_x^2}\right)\left(1-\sqrt{1-\frac{i_{x1}^2}{c_{x1}^2}}\right); \text{ to inflexion}$$

$$\searrow n_x\frac{dmc_x^2}{dt_i\varepsilon_{x-}}\left(1-\sqrt{1-\frac{(c_x-i_x)^2}{c_x^2}}\right); \text{ for gravitation}$$

3A8
$$\frac{dmc_{x1}^2}{dt_{op}}\left(1-\sqrt{1-\frac{i_{x1}^2}{c_{x1}^2}}\right) \rightarrow \frac{dmc_{x1}^2}{dt_{on1}}\sqrt{1-\frac{(c_{x1}-i_{x1})^2}{c_{x1}^2}}\left(1-\sqrt{1-\frac{i_{x1*}^2}{c_{x1*}^2}}\right)$$

3A9
$$\frac{dmc_{x1}^2}{dt_{on1-}}\sqrt{1-\frac{(c_{x1}-i_{x1})^2}{c_{x1}^2}}\left(1-\sqrt{1-\frac{i_{x1*}^2}{c_{x1*}^2}}\right) \rightarrow \frac{dmc_{x1}^2}{dt_{op1-}}\left(1-\frac{(c_{x1}-i_{x1})^2}{c_{x1}^2}\right)\left(1-\sqrt{1-\frac{i_{x2}^2}{c_{x2}^2}}\right); \text{ to infl.}$$

$$\searrow n_{x1}\frac{dmc_{x1}^2}{dt_i\varepsilon_{x1-}}\left(1-\sqrt{1-\frac{(c_{x1}-i_{x1})^2}{c_{x1}^2}}\right); \text{ for gravitation}$$

Diag. 3.1 Diagram 3.1

As
- the intensities of the anti-electron process quantum drives are equal, and
- the cover of the anti-neutron process is the one establishing the proton process,

the intensity of the proton process corresponds to the intensity of the cover, which value in fact depends only from the speed value of the quantum communication.

Diagram 3.1 on the previous page demonstrates the case:
1. The anti-process generates the anti-proton collapse, the intensity of which establishes the intensity of the proton process. The higher the quantum speed is, the more massive the quantum impact is, the higher is the frequency of the start of the elementary process, the higher is the intensity of the start of the proton process.
2. The higher the quantum speed of the anti-electron process is, the longer is it, as drive in action.
3. The anti-proton collapse, with reference to 1B11, is with the decrease of the quantum speed and with the increase of the time count, because of the generation of the quantum impulse. (Quantum speed c_{x1} in 3A6 corresponds already to the modified value in 3A7, the same way as c_{x1*} and c_{x2} are in 3A7 and 3A8 as well.)

Ref. 1B11

Acceleration in conventional terms means the speeding up of the mass/weight, measured on the *Earth* surface. In the case of elementary processes, with reference to Section 2.2, it means the increase or the decrease of the number of the quantum, impacted by the elementary process.
This might be misunderstood, having our conventional view in use.
This is not about the impact of the surrounding environment by the surface of the accelerating subject! (Like higher speed results in more physical contact.) This is about the "internal quantum impact" of infinite number of elementary cycles, running in parallel.

Ref. S. 2.2

Acceleration, driven by internal source reduces the potential of the process; external, increases it. This is not just about part of the elementary process. This is a modification, having immediate impact on the whole process, independently on the stage of its progress:

The electron process and the anti-electron process drives are always in action. There is no collapse without drive, and there is no transfer demand without collapse. In the case of any external impact, which is increasing or decreasing the *IQ* quantum drive of the electron process, the effect is immediate.

Solid, liquid or gaseous statuses do not make any difference. The increase or the decrease of the number of the impacted quantum happens anyway.
The difference between the solid and the other two statuses is that the solid status is "originally" without internal conflict. (It might be different after external acceleration, as it increases the internal intensity.) The acceleration has its immediate impact. The speeding up however does not change the intensity relations of the elementary processes. The elementary process remains the same. The number of the impacted quantum is the one, which becomes increased or decreased, function of the energy source of the acceleration.

S. 3.2

3.2
The sequence of elementary cycles

Quantum drive $\frac{dmc_x^2}{dt_i \varepsilon_x}$ means infinite variety of quantum speed and intensity values.

Ref. 1B11

Elementary processes always remain the same, with the natural modification of the quantum speed value during the elementary cycle.

3B1 $\varepsilon_{x-} = \frac{\varepsilon_{n-}}{\varepsilon_{p-}}$; the intensities of the anti-processes are reciprocal: $\varepsilon_{x-} = \frac{1}{\varepsilon_x}$. Why reciprocal?

The time count of the status of rest is equal and means $\lim \Delta t = 0$ for all elementary processes.

The time count of the proton and the neutron processes depends on the intensity relation of the proton and the neutron processes. But this is in fact a relation, where the intensity of the neutron process (in conventional terms: the mass) in each elementary case is related to the intensity of the proton process.

3B2	This way, the intensity of the electron process is expressing in conventional terms the relation of the masses of the protons and the neutrons of the element.	$\varepsilon_e = \frac{\varepsilon_p}{\varepsilon_n}\sqrt{1-\frac{(c_x - i_x)^2}{c_x^2}}$;
3B3	The meaning of the intensity of the electron process in conventional terms therefore is:	$\varepsilon_e = \frac{\sum m_p}{\sum m_n} = \frac{n_x \cdot m_p}{m_x - (n_x \cdot m_p + n_x \cdot m_e)}$;
3B4	The equation in 3B2 can also be written: (In this case $\varepsilon_n \neq \varepsilon_{neutron}$, but the value of ε_e is the same.)	$\varepsilon_e = \frac{1}{\varepsilon_{neutron}}\sqrt{1-\frac{(c_x - i_x)^2}{c_x^2}}$;

The intensity of the anti-neutron process is repeating the intensity of the collapse of the neutron process at the inflexion. The intensities of the drives of the anti-proton processes are equal, just of different durations and intensities of the inflexion. (In line with the value of the quantum speed and the number of the elementary cycles of the process).

The reciprocal character is valid and we never use the exact intensity values of the proton and neutron processes, just the intensities of the electron and anti-electron processes.

In the case of *neutron process dominance*, the intensity of the drive is result of the intensity "pressure" of the quantum membrane of the anti-electron process surplus. This is the reason of the increased intensity of the electron process.

In the case of *proton process dominance*, the electron process surplus is formulating on the direct side. This way the quantum membrane is generating on the direct side and the anti-electron process is the one, having its increased intensity: $\varepsilon_x > 1$ and $\varepsilon_{x-} < 1$.

The neutron processes of *proton process dominant elementary* processes can easily be driven from aside, because the intensity of their electron process drives is of lesser value. *Neutron process dominant elementary processes* usually have increased values of electron process intensity.

In the case, electron processes drive neutron collapse, belonging to other elementary processes, their neutron processes become free to be driven from aside.
Proton process dominant elementary processes produce surplus in drives, but they drive neutron processes with lesser intensities. This means they offer their neutron processes to be driven from aside by elementary processes with higher intensities. Higher intensities take over lesser ones and the anti-drives will be producing proton processes with higher quantum speed and higher intensities, corresponding to the parameters of the drive, the neutron process dominant elementary processes.

With the increase of the intensity of the neutron process dominant elementary processes by acceleration, the strength of the quantum membrane at the anti-electron process side is increasing. It results an inflexion of increased intensity.

Anti-electron process: | Anti-electron IQ drive:

$$e_{ea} = \frac{dmc_x^2}{dt_i\varepsilon_x}\sqrt{1-\frac{(c_x-i_x)^2}{c_x^2}}\left(\frac{1}{\sqrt{1-\frac{v^2}{c_x^2}}}-1\right)\left(1-\sqrt{1-\frac{(c_x-i_x)^2}{c_x^2}}\right); \quad IQ_{xa} = \frac{c_x^2\sqrt{1-\frac{(c_x-i_x)^2}{c_x^2}}}{\varepsilon_x\sqrt{1-\frac{v^2}{c_x^2}}}; \qquad \text{3B5}$$

Elementary processes with *proton process dominance* are the initiators of the communication, since they are the ones offering their neutron processes to be driven.

Neutron process dominant elementary processes have increased electron process intensity, result of the increased quantum membrane of the surplus of the anti-electron processes.
In the case of conflicts, elementary processes communicate in order to resolve the conflict.

3.2.1 S. 3.3.1

In the case of ***proton process dominant*** elementary processes, the intensity of the proton process is higher than the intensity of the driven neutron collapse.

$$\frac{dmc^2}{dt_p} > \frac{dmc^2}{dt_n}; \text{ as } \varepsilon_p > \varepsilon_n \quad \text{as } \varepsilon = \frac{1}{\Delta t} \quad \text{and } \varepsilon_e > 1 \qquad \text{3C1}$$

if for the equal unit period of time
the number of the generating proton processes is n_p and
the number of the driven neutron process is n_n
| the relation is $n_p > n_n$ 3C2

The electron process surplus is generating and accumulating on the direct side of the elementary process.
The cycle is being established by the neutron process, as $\varepsilon_{n-} = \varepsilon_n$;

$$\text{this way: } \frac{dmc^2}{dt_{n-}} < \frac{dmc^2}{dt_{p-}}; \qquad \text{3C3}$$

If the number of the acting cycles corresponds to n_n the number of the proton processes of the next cycle will be: $n_p^{n+1} = n_{p-}^n = n_{n-}^n = n_n^n < n_p^n$;

3C5 Meaning: $n_p^{n+1} < n_p^n$

The number of the elementary processes in absolute terms is decreasing.
With reference to 3C5, the equation above: $n_{p-}^n = n_{n-}^n$ because $\varepsilon_{e-} < 0$

There are accumulating electron and proton process intensities, while the number of the elementary processes is decreasing. For keeping the stability of the elementary cycles, the elementary communication for proton process dominant elementary processes is vital.

S. 3.2.2

3.2.2

In the case of ***neutron process dominant*** elementary processes, the intensity of the electron process is established by the accumulating surplus of the anti-electron processes. As the intensities of all anti-electron process drives are identical, the surplus of the anti-electron processes is released. This way the quantum impact and the number of cycles would be less and less, but the regeneration of the surplus is continuous.

3D1 $$\frac{dmc^2}{dt_p} < \frac{dmc^2}{dt_n}; \quad \text{as } \varepsilon_p < \varepsilon_n \text{ and } \varepsilon = \frac{1}{\Delta t} \quad \text{and } \varepsilon_e < 1, \text{ as } \frac{\varepsilon_p}{\varepsilon_n} < 1$$

As the intensity of the neutron process is higher, there is no surplus neither in the number of the proton, neither in the number of the electron processes. The intensity of the electron process as drive shall be of increased value and all electron processes shall be functioning as drive: Therefore $n_n = n_p$

The intensities of the neutron/anti-neutron processes $\varepsilon_{n-} = \varepsilon_n$ and

3D2 $$\text{this way: } \frac{dmc^2}{dt_{n-}} > \frac{dmc^2}{dt_{p-}}; \text{ which means: } \varepsilon_{e-} > 1, \text{ as } \frac{\varepsilon_{n-}}{\varepsilon_{p-}} = \varepsilon_{e-}$$

There is a surplus at the anti-electron process side, as the intensities of the neutron/anti-neutron processes are higher than the intensities of the anti-proton/proton processes.
This way the operating cycle will be: $n_p^{n+1} = n_{p-}^n < n_{n-}^n = n_n^n = n_p^n$;

3D3 Meaning: $n_p^{n+1} < n_p^n$

This is decreasing tendency as well, similarly to the proton process dominant case.
For establishing the overall balance, the quantum communication for neutron process elementary processes is vital.

S. 3.2.3

3.2.3

All elementary processes have different quantum speed values therefore **each elementary process has its own space-time**.

Ref. 1A5 1A7

The quark processes and others, representing the proton process happen in identical ways in any elementary systems – just the space-times and the durations are different. With reference to 1A5-1A7 the proton process means a certain number of quantum impulses, impacted. The intensities and the durations of the proton processes space-time by space-time are different.

The intensity of the proton process is: $\frac{dmc_x^2}{dt_p}\left(1-\sqrt{1-\frac{v^2}{c_x^2}}\right)=\frac{dmc_x^2}{dt_o}\sqrt{1-\frac{v^2}{c_x^2}}\left(1-\sqrt{1-\frac{v^2}{c_x^2}}\right)$; 3E1

and the quantum speed value generates the time count.

$$dt_{x1}=\frac{dt_o}{\sqrt{1-\frac{v^2}{c_{x1}^2}}};\quad dt_{x2}=\frac{dt_o}{\sqrt{1-\frac{v^2}{c_{x2}^2}}};$$

The time counts are different, but the event (v) in both systems are synchronised. 3E2

If $c_{x1}>c_{x2}$ and $dt_{x1}<dt_{x2}$, and $\frac{c_{x1}^2}{\Delta t_{x1}}>\frac{c_{x2}^2}{\Delta t_{x2}}$; 3E3

Measuring the event within the time system of the *Earth*, the speeding up to v within the space-time of the *Earth* results in each and all cases identical time durations.

$$dt_{Earth}=\frac{dt_o}{\sqrt{1-\frac{v^2}{c_{Earth}^2}}}$$ 3E4

The drive of the anti-proton processes are equal, as

$$\frac{dmc_x^2}{dt_i\varepsilon_{x-}}\left(1-\sqrt{1-\frac{(c_x-i_x)^2}{c_x^2}}\right)=const;\text{ as }\frac{c_x^2}{\varepsilon_{x-}}=const$$

the time system here does not have it distinguishing character; dt_i for all electron processes are the same. 3E5

The intensities of the quantum drives of all anti-proton collapses are identical, with the infinite large variety of the durations of the collapse and the intensities of the inflexion. The intensities of the proton processes vary.

The elementary cycle shall be continuous, independently on the number of the cycles of the elementary process and the intensities of the electron processes. There should be no "gap" between the developing neutron processes and the covering proton processes. The drive and the cover shall be guaranteed, whatever the intensity value is. The key of the continuity of the process is the fluent sequence of the elementary cycles.

The developing conflicts of the surplus either on the anti-, or on the direct sides are equivalent to the increased number of electron or anti-electron processes. The release of the anti-electron processes as quantum impacts of *gravitation* and the use of electron process surplus in elementary communication means the decrease of the elementary cycles. (The intensity surplus of the anti-neutron process generates the expansion, the mechanical impact of gravitation; the proton process surplus on the direct side is the potential of the communication.) The generation of the surplus is obvious, as the intensities of the proton and neutron processes are different.

3.3 The meaning of the *parallel* and the *internal* elementary cycles

S. 3.3

The Periodic Number of the elementary processes corresponds to the number of the proton processes. This also means the number of the elementary cycles.
Periodic Number n means there are n cycles within the elementary process running in parallel.

The number of the cycles operating in parallel is a kind of additional indicator to the intensity of the process.
The higher the periodic number is,
- the higher is the intensity of the quantum drive of the electron process;
- the higher are the intensities of the neutron and the anti-neutron processes;
- the higher are the proportions of the generation of the anti-electron processes against their use as anti-proton process drive;
- the higher is the surplus of the anti-electron processes!

Each parallel elementary cycle is result of the anti-electron process drive. The Periodic Number is the initial number of the generating anti-proton and proton processes.
The less the Periodic Number is,
- the less is the intensity of the quantum drive of the electron process;
- the less is the number of the impacts of the anti-electron processes!

The *Uranium* process, for example, is "loosing" more than one third of its anti-electron process drives by the generating surplus.
The process starts with cycle number n_{U1} of the proton processes; the end of the direct process is n_{U1} neutron processes as well. The anti-direction starts by the same n_{U1} number of anti-neutron processes, but there is generation of anti-electron process surplus and conflict in the anti-direction of the cycle. The generating n_{U1} number of the anti-electron processes includes this surplus as well.

The new elementary cycle starts with quasi the same quantum speed, but with proton processes, less in numbers. While the anti-electron processes are in surplus, their number in drives is significantly less.

3F1 The surplus of the anti-electron process drives is $\Delta n_U = (\varepsilon_{U-} - 1)n_{U1}$.

The value of the intensity coefficient of the drive of the anti-proton processes – as it is of standard *IQ* value – significantly less (ε_{x-} the intensity coefficient of the anti-electron process is of higher value).

As intensity means events for the unit period of time, we can also formulate this, as the less generation of the proton processes in numbers than the number of the expanding anti-neutron processes, for the unit period of time.
The less need in intensity is equal to the intensity of the generation multiplied by the

3F2 "missing" umbers: $\Delta n_{Ux(x+1)} \cdot \varepsilon_{U-} = (\varepsilon_{U-} - 1)n_{Ux}$

The formulating number of anti-proton/proton inflexions(for the next cycle) is:

3F3
$$n_{U2} = n_{U1} - \frac{\varepsilon_{U-} - 1}{\varepsilon_{U-}} n_{U1} = n_{U1}\left(1 - \frac{\varepsilon_{U-} - 1}{\varepsilon_{U-}}\right);$$

3F4 $c_x^2 \cdot \varepsilon_x = \dfrac{c_x^2}{\varepsilon_{x-}} = const$ the surplus means: $\varepsilon_{x-} = \dfrac{\varepsilon_{n-}}{\varepsilon_{p-}}$ | $\varepsilon_{x-} > 1$ means, $(\varepsilon_{x-} - 1)$ proportion has been "lost"

The generating anti-electron process surplus does not mean at all the accumulation of anti-electron processes, remaining without use:

(1) the conflict is establishing the intensity of the electron process on the direct side – as consequence of the conflicting tense of the internal quantum membrane;

(2) at the same time, the *blue shift* impact of the anti-electron process is the quantum impact of gravitation.

The higher the Periodic Number is, the higher is the surplus of the anti-electron processes; the higher is the quantum impact of gravitation; the higher is the intensity of the electron process, generated by the internal quantum membrane!

Without the necessary high number of parallel cycles, the elementary process would be completed shortly. This is the reason of the high periodic number of the *Uranium* process in the example. The low proportion of the *internal cycles* needs high number of *parallel cycles*!

The *Uranium* process is in constant operation (as all elementary processes are) with permanent generation of the surplus and the quantum impact of *gravitation.*

Table 3.1 shows those 92 standard elementary cycles of the *Uranium* process become reduced to 58 after the first neutron/anti-neutron inflexion. Without self-rehabilitating steps the cycles would in practice expire after the 7th elementary inflexion.

Self-rehabilitation means the integrated operation of infinite large number of *Uranium* processes. This way the elementary cycles are connected to each other and operate hand in hand.

For proton process dominance, we can take the *Oxygen* process as example. The reduction of the cycles in this case happens in the direct process, as $\varepsilon_O > 1$.

The intensity of the anti-electron process, for the *Oxygen* processes is $\varepsilon_{O-} < 1$, therefore there is equal number of anti-proton and anti-neutron processes on the anti-direction

$$n_{Op-} = n_{On-}. \qquad \text{3F5}$$

Therefore the reduction is on the direct side and the number of the neutron collapses is establishing the number of the parallel elementary cycles of the next cycle:

$$n_{O2} = n_{O1} - \frac{\varepsilon_O - 1}{\varepsilon_O} n_{O1} = n_{O1}\left(1 - \frac{\varepsilon_O - 1}{\varepsilon_O}\right) \qquad \text{3F6}$$

All elementary processes exist and manage infinite number of elementary processes, with the necessary number of parallel cycles for their stable operation. Once the number of the operating parallel cycles cannot be guaranteed, elementary evolution steps ahead to the next sequence to the new elementary process with less number of proton processes.

With reference to 3F3 and 3F5, the general formula of the change of the parallel cycles is function of the number of the previous cycle and the intensities of the electron and the anti-electron processes:

3F7 $$n_{n+1} = n_n\left(1 - \frac{\varepsilon_{x-} - 1}{\varepsilon_{x-}}\right);$$ for neutron process dominance, if $\varepsilon_{x-} > 1$; $\left(\left(1 - \frac{\varepsilon_{x-} - 1}{\varepsilon_{x-}} = \varepsilon_x\right)\right)$

3F8 $$n_{n+1} = n_n\left(1 - \frac{\varepsilon_x - 1}{\varepsilon_x}\right);$$ for proton process dominance, if $\varepsilon_x > 1$; $\left(\left(1 - \frac{\varepsilon_x - 1}{\varepsilon_x} = \varepsilon_{x-}\right)\right)$

For proton process dominant elementary processes the surplus is used *for elementary communication.* For neutron process dominant elementary processes the surplus is the source of the *quantum impact of gravitation.*

The following Table 3.1 shows the number of the parallel cycles of the elementary processes of the Periodic Table for the first 7 cycles.

The table shows that the parallel cycles of the elementary processes with neutron process dominance are consolidating to a general common number, while the elementary processes with proton process dominance mainly keep their numbers.

Elementary processes close to equilibrium keep their parallel cycles stable. The best example is the *Neon* process with its $\varepsilon_{Ni} = 1.002$ value of the intensity coefficient.

Proton process dominant elementary processes remain active for long, as the gradient of the reduction of the number of their parallel cycles is very low value.

The initial number of the parallel cycles in Table 3.1 below corresponds to the Periodic Number. The reduction can be followed by the columns with the numbers of the sequences of the elementary process.

Elementary process	**PN**	mass	ε_{e-}	ε_e	**1**	**2**	**3**	**4**	**5**	**6**	**7**
Hydrogen	1	1.008									
Helium	2	4.003	0.986	1.014	***2.0***	***1.9***	***1.9***	***1.9***	***1.9***	***1.8***	***1.8***
Lithium	3	6.940	1.296	0.772	**2.3**	**1.8**	**1.4**	**1.1**	**0.8**	**0.6**	**0.5**
Beryllium	4	9.012	1.236	0.809	**3.2**	**2.6**	**2.1**	**1.7**	**1.4**	**1.1**	**0.9**
Boron	5	10.810	1.146	0.873	**4.4**	**3.8**	**3.3**	**2.9**	**2.5**	**2.2**	**1.9**
Carbon	6	12.011	0.987	1.013	***5.9***	***5.8***	***5.8***	***5.7***	***5.6***	***5.5***	***5.5***
Nitrogen	7	14.007	0.986	1.014	***6.9***	***6.8***	***6.7***	***6.6***	***6.5***	***6.4***	***6.3***
Oxygen	8	15.999	0.985	1.015	***7.9***	***7.8***	***7.6***	***7.5***	***7.4***	***7.3***	***7.2***
Fluorine	9	18.998	1.095	0.913	**8.2**	**7.5**	**6.9**	**6.3**	**5.7**	**5.2**	**4.8**
Neon	10	20.170	1.002	0.998	**10.0**	**10.0**	**9.9**	**9.9**	**9.9**	**9.9**	**9.9**
Sodium (Na)	11	22.989	1.074	0.931	**10.2**	**9.5**	**8.9**	**8.3**	**7.7**	**7.2**	**6.7**
Magnesium	12	24.305	1.010	0.990	**11.9**	**11.8**	**11.6**	**11.5**	**11.4**	**11.3**	**11.2**
Aluminium	13	26.982	1.056	0.947	**12.3**	**11.7**	**11.0**	**10.5**	**9.9**	**9.4**	**8.9**
Silicon	14	28.086	0.991	1.009	***13.9***	***13.8***	***13.6***	***13.5***	***13.4***	***13.3***	***13.2***
Phosphorus	15	30.974	1.050	0.953	**14.3**	**13.6**	**13.0**	**12.4**	**11.8**	**11.2**	**10.7**
Sulphur	16	32.060	0.989	1.011	***15.8***	***15.6***	***15.5***	***15.3***	***15.1***	***14.9***	***14.8***
Chlorine	17	35.453	1.070	0.935	**15.9**	**14.9**	**13.9**	**13.0**	**12.1**	**11.3**	**10.6**
Argon	18	39.948	1.203	0.831	**15.0**	**12.4**	**10.3**	**8.6**	**7.1**	**5.9**	**4.9**

Potassium (K)	19	39.098	1.042	0.959	18.2	17.5	16.8	16.1	15.4	14.8	14.2
Calcium	20	40.080	0.989	1.011	19.8	19.6	19.3	19.1	18.9	18.7	18.5
Scandium	21	44.956	1.125	0.889	18.7	16.6	14.8	13.1	11.7	10.4	9.2
Titanium	22	47.900	1.161	0.861	18.9	16.3	14.1	12.1	10.4	9.0	7.7
Vanadium	23	50.942	1.198	0.835	19.2	16.0	13.4	11.2	9.3	7.8	6.5
Chromium	24	51.996	1.150	0.869	20.9	18.1	15.8	13.7	11.9	10.4	9.0
Manganese	25	54.938	1.181	0.847	21.2	17.9	15.2	12.8	10.9	9.2	7.8
Iron (Fe)	26	55.847	1.132	0.883	23.0	20.3	17.9	15.8	14.0	12.4	10.9
Cobalt	27	58.933	1.166	0.857	23.1	19.8	17.0	14.6	12.5	10.7	9.2
Nickel	28	58.710	1.081	0.925	25.9	24.0	22.2	20.5	19.0	17.5	16.2
Cuprum	29	63.540	1.175	0.851	24.7	21.0	17.9	15.2	13.0	11.0	9.4
Zinc	30	65.380	1.163	0.860	25.8	22.2	19.1	16.4	14.1	12.1	10.4
Gallium	31	69.735	1.233	0.811	25.1	20.4	16.5	13.4	10.9	8.8	7.2
Germanium	32	72.590	1.252	0.799	25.6	20.4	16.3	13.0	10.4	8.3	6.7
Arsenic	33	74.922	1.253	0.798	26.3	21.0	16.8	13.4	10.7	8.5	6.8
Selenium	34	78.960	1.305	0.766	26.1	20.0	15.3	11.7	9.0	6.9	5.3
Bromine	35	79.904	1.266	0.790	27.6	21.8	17.3	13.6	10.8	8.5	6.7
Krypton	36	83.800	1.310	0.763	27.5	21.0	16.0	12.2	9.3	7.1	5.4
Rubidium	37	85.466	1.293	0.774	28.6	22.1	17.1	13.2	10.2	7.9	6.1
Strontium	38	87.620	1.289	0.776	29.5	22.9	17.8	13.8	10.7	8.3	6.4
Yttrium	39	88.906	1.263	0.792	30.9	24.5	19.4	15.3	12.2	9.6	7.6
Zirconium	40	91.220	1.263	0.792	31.7	25.1	19.8	15.7	12.4	9.8	7.8
Niobium	41	92.906	1.249	0.801	32.8	26.3	21.0	16.8	13.5	10.8	8.6
Molybdenum	42	95.940	1.267	0.789	33.1	26.2	20.6	16.3	12.9	10.1	8.0
Technetium	43	98.962	1.284	0.779	33.5	26.1	20.3	15.8	12.3	9.6	7.5
Ruthenium	44	101.070	1.280	0.781	34.4	26.9	21.0	16.4	12.8	10.0	7.8
Rhodium	45	102.906	1.270	0.788	35.4	27.9	22.0	17.3	13.6	10.7	8.5
Palladium	46	106.400	1.296	0.772	35.5	27.4	21.1	16.3	12.6	9.7	7.5
Silver	47	107.868	1.278	0.783	36.8	28.8	22.5	17.6	13.8	10.8	8.4
Cadmium	48	112.410	1.324	0.755	36.2	27.4	20.7	15.6	11.8	8.9	6.7
Indium	49	114.820	1.326	0.754	37.0	27.9	21.0	15.9	12.0	9.0	6.8
Tin	50	118.690	1.356	0.737	36.9	27.2	20.0	14.8	10.9	8.0	5.9
Antimony	51	121.750	1.370	0.730	37.2	27.2	19.9	14.5	10.6	7.7	5.6
Tellurium	52	127.600	1.436	0.697	36.2	25.2	17.6	12.2	8.5	5.9	4.1
Iodine	53	126.905	1.377	0.726	38.5	28.0	20.3	14.8	10.7	7.8	5.7
Xenon	54	131.300	1.413	0.708	38.2	27.0	19.1	13.5	9.6	6.8	4.8
Caesium	55	132.905	1.399	0.715	39.3	28.1	20.1	14.4	10.3	7.4	5.3
Barium	56	137.330	1.434	0.697	39.0	27.2	19.0	13.2	9.2	6.4	4.5
Lanthanum	57	138.906	1.419	0.705	40.2	28.3	20.0	14.1	9.9	7.0	4.9
Cerium	58	140.120	1.398	0.715	41.5	29.7	21.2	15.2	10.9	7.8	5.6
Praseodymium	59	140.908	1.371	0.730	43.0	31.4	22.9	16.7	12.2	8.9	6.5
Neodymium	60	144.240	1.381	0.724	43.4	31.5	22.8	16.5	11.9	8.6	6.3
Promethium	61	145.000	1.359	0.736	44.9	33.0	24.3	17.9	13.1	9.7	7.1

Samarium	62	150.400	1.408	0.710	44.0	31.3	22.2	15.8	11.2	8.0	5.7
Europium	63	151.960	1.394	0.717	45.2	32.4	23.2	16.7	12.0	8.6	6.2
Gadolinium	64	157.250	1.439	0.695	44.5	30.9	21.5	14.9	10.4	7.2	5.0
Terbium	65	158.925	1.427	0.701	45.6	31.9	22.4	15.7	11.0	7.7	5.4
Dysprosium	66	162.500	1.444	0.693	45.7	31.7	21.9	15.2	10.5	7.3	5.0
Holmium	67	164.930	1.443	0.693	46.4	32.2	22.3	15.4	10.7	7.4	5.1
Erbium	68	167.260	1.441	0.694	47.2	32.7	22.7	15.8	10.9	7.6	5.3
Thulium	69	168.934	1.430	0.699	48.2	33.7	23.6	16.5	11.5	8.1	5.6
Ytterbium	70	173.040	1.454	0.688	48.2	33.1	22.8	15.7	10.8	7.4	5.1
Lutetium	71	174.967	1.446	0.692	49.1	34.0	23.5	16.2	11.2	7.8	5.4
Hafnium	72	178.490	1.461	0.685	49.3	33.7	23.1	15.8	10.8	7.4	5.1
Tantalum	73	180.948	1.460	0.685	50.0	34.2	23.4	16.1	11.0	7.5	5.2
Tungsten	74	183.850	1.466	0.682	50.5	34.4	23.5	16.0	10.9	7.5	5.1
Rhenium	75	186.207	1.464	0.683	51.2	35.0	23.9	16.3	11.1	7.6	5.2
Osmium	76	190.200	1.484	0.674	51.2	34.5	23.3	15.7	10.6	7.1	4.8
Iridium	77	192.220	1.478	0.677	52.1	35.3	23.9	16.1	10.9	7.4	5.0
Platinum	78	195.090	1.483	0.674	52.6	35.5	23.9	16.1	10.9	7.3	5.0
Gold	79	196.967	1.475	0.678	53.6	36.3	24.6	16.7	11.3	7.7	5.2
Mercury	80	200.590	1.489	0.672	53.7	36.1	24.2	16.3	10.9	7.3	4.9
Thallium	81	204.370	1.504	0.665	53.8	35.8	23.8	15.8	10.5	7.0	4.6
Lead	82	207.800	1.515	0.660	54.1	35.7	23.6	15.6	10.3	6.8	4.5
Bismuth	83	208.980	1.499	0.667	55.4	36.9	24.6	16.4	11.0	7.3	4.9
Polonium	84	*209.00*	1.470	0.681	57.2	38.9	26.5	18.0	12.3	8.3	5.7
Astatine	85	*210.00*	1.452	0.689	58.5	40.3	27.8	19.1	13.2	9.1	6.2
Radon	86	*222.00*	1.562	0.640	55.1	35.2	22.6	14.4	9.2	5.9	3.8
Francium	87	*223.00*	1.542	0.649	56.4	36.6	23.7	15.4	10.0	6.5	4.2
Radium	88	226.025	1.549	0.645	56.8	36.7	23.7	15.3	9.9	6.4	4.1
Actinium	89	*227.00*	1.532	0.653	58.1	37.9	24.8	16.2	10.6	6.9	4.5
Thorium	90	232.038	1.559	0.641	57.7	37.0	23.8	15.2	9.8	6.3	4.0
Protactinium	91	231.036	1.520	0.658	59.9	39.4	25.9	17.0	11.2	7.4	4.9
Uranium	92	238.029	1.568	0.638	58.7	37.4	23.9	15.2	9.7	6.2	3.9
Neptunium	93	237.048	1.530	0.654	60.8	39.7	26.0	17.0	11.1	7.2	4.7
Plutonium	94	*244.00*	1.577	0.634	59.6	37.8	24.0	15.2	9.7	6.1	3.9

Table 3.1

Table 3.1

Diagrams 3.1 and 3.2 on the following page are presenting the change of the operating in parallel elementary cycles in line with 21 elementary sequences. Diagram 3.2 shows the gradient of the change of all elementary processes; Diagram 3.3 is only about the first 29 elementary processes from the *Helium* process, up to the *Cuprum* process.

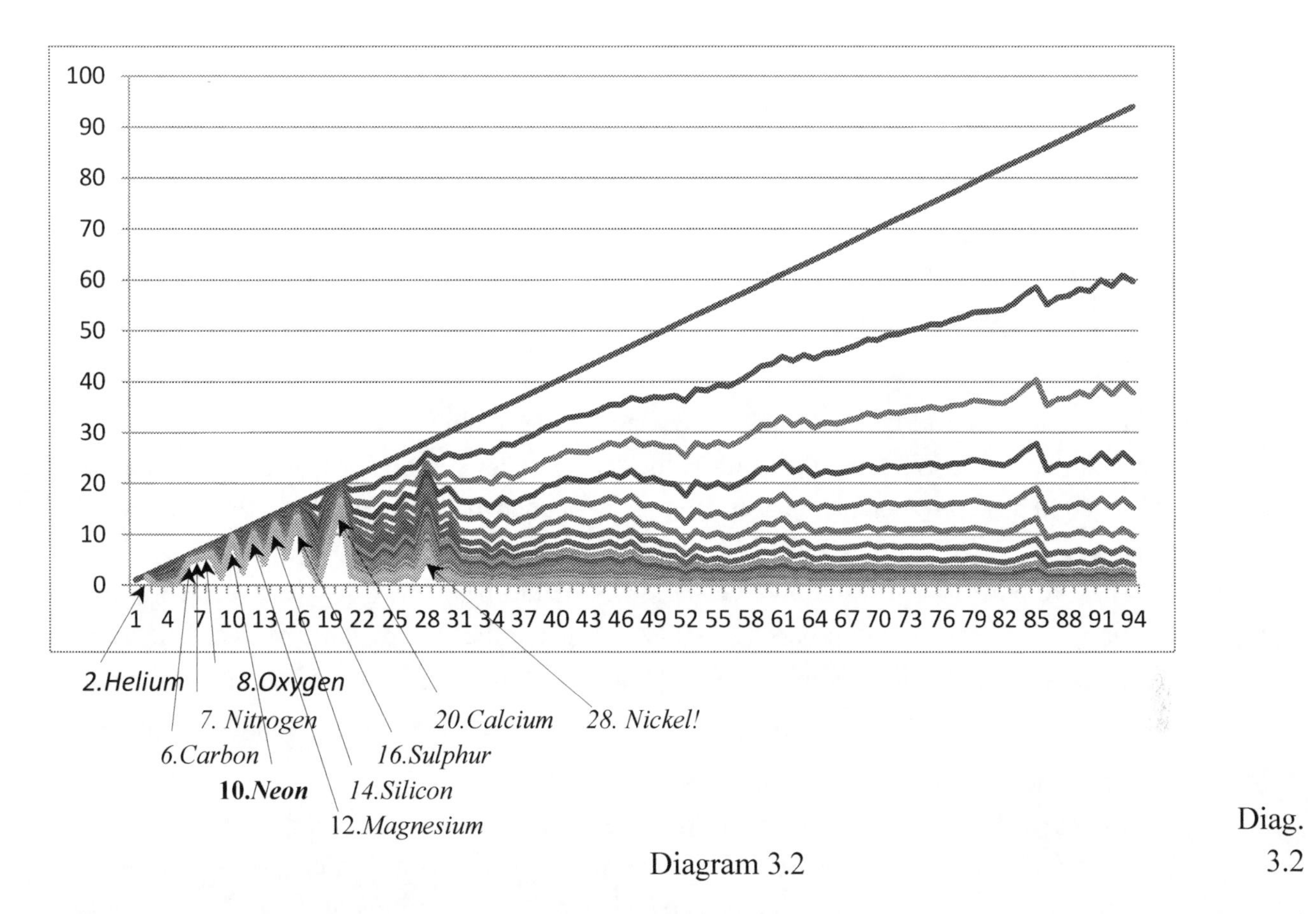

Diagram 3.2

Diag. 3.2

Diagram 3.3 introduces those elementary processes, which are less sensitive to the elementary progress. Elementary processes up to the *Cuprum* process, *No 29* show variety of impacts.

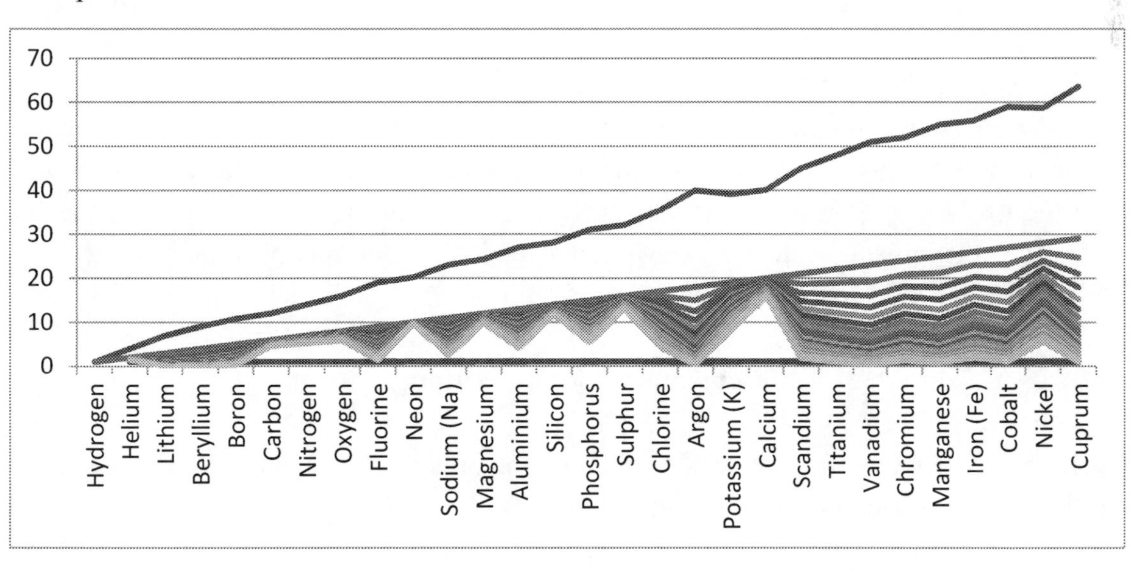

Diag. 3.3

Diagram 3.3

Both diagrams above show the first 21 consecutive double inflexion sequences of the elementary processes.

Elementary cycles mean the running in parallel elementary processes.

- proton process dominant elementary processes modify their original status slowly;
- neutron process dominant elementary processes, with high intensity electron process drive, have increased gradient of the elementary change; without the operating parallel cycles the process would be quickly over.

One of the lessons of the diagrams above is the unique, close to the equilibrium status of the *Neon* process.

The *Lithium*, *Beryllium* and *Boron* processes also have their specifics, being at the very beginning of the Periodic Table and at the same time having so heavily increased electron process intensity, corresponding in its value to elementary processes of periodic numbers around 40. These three elementary processes have their relating importance in the nuclear industry.

These elementary processes have increased neutron process dominance, at the same time, with very few numbers (3, 4 and 5) of parallel elementary cycles. The *Boron* process for example is with $\varepsilon_{B-} = 1.146$ and with initial number of elementary cycles: $n_{Bo} = 5$. For managing a more balanced operation, their cycle reduction intensity would need a higher number of elementary cycles.

Boronic-acid is used in pressurised water reactors for "so called neutron capture" function at the starting phases of the operation. The reason is that the *Boron* and the *Uranium* processes communicate. The electron process surplus of the *U-235* content of the nuclear fuel drives the neutron processes of the *Boron* processes and vice versa. The integrated result of the communication improves the ration of the working in parallel elementary cycles of the *Boron* process, as the *Uranium* process is initiating the taking over driving functions from the *Boron* process.

The key is the high ration of the loss of the *Boron* cycles. With the drive, coming from the *Uranium* electron process surplus (the reason in fact of the fission) – which is the dominant one in the communication (as it is of the higher intensity) the absolute value of the loss becomes reduced. At the same time the electron process surplus of the *U-235* fuel, the initiator of the nuclear fission, expires in the communication.

S. 3.4

3.4 Elementary communication

The difference between proton dominant and neutron dominant elementary processes gives the rational of the quantum communication:

As the number of the elementary cycles in proton process dominant elementary processes depends on the direct side, the more of them is used in communication, the less is the loss. At the same time the more neutron process dominant elementary processes are in communication with proton process dominant elementary processes, the less is the portion, lost in the anti-direction.

Elementary communication needs ***electron process conflict***.
The conflict is always generating on the direct side. Anti-direction is for the self-control of the elementary process, with the intensity consequence of the anti-electron process and the release of the anti-electron process impact.
The higher is the conflict, the higher is the intensity of the communication.

The initiation of the conflict is coming from proton process dominant elementary processes, especially from the *Hydrogen*, *Helium*, *Nitrogen* and the *Oxygen* processes.

Water has its specific role as "conflict generator".
The other four proton process dominant elementary processes have their specific compositions as well: The *Carbon* process together with the *Hydrogen* process is acting as *Hydrocarbons*; the *Calcium*, the *Sulphur* and the *Silicon* especially with their *Oxides*, and in combination with other elementary processes as *Carbonates*, *Sulphides*, *Silicates, and Calcites*.

The most usual and the simplest way of the generation of conflict is the use of *water* and *hydrocarbons*.
The key in the generation of the conflict is the *Hydrogen* process.

S. 3.4.1

3.4.1.

Water is product of the nature. Its generation needs the conflict of the *Oxygen* and the *Hydrogen* processes. This conflict is born as part of the elementary evolution.
The original speed of the quantum communication of the *Oxygen* process is less than the quantum speed on the *Earth* surface. The speed of the quantum communication of the *Hydrogen* process is also less; it is of infinite low value.

The original conflict of these elementary processes with the permanent quantum impact of *Earth gravitation* increases the quantum speed of these elementary processes equalling them to the quantum speed of *Earth* gravitation in their gaseous state.

The *Oxygen* process, with its electron process *blue shift* surplus, as product of elementary evolution, drives all available electron/neutron processes of the *Hydrogen* process. This way the intensity potential of the proton process of the *Hydrogen* process is accumulating.

$$\frac{dmc_O^2}{dt_i\varepsilon_O}\left(1-\sqrt{1-\frac{(c_O-i_O)^2}{c_O^2}}\right);\ \varepsilon_O=1.01533 \text{ and } \frac{dmc_H^2}{dt_i\varepsilon_H}\left(1-\sqrt{1-\frac{(c_H-i_H)^2}{c_H^2}}\right);\ \lim\varepsilon_H=\infty \qquad \text{3G1}$$

The entropy product of the electron process (where the neutron collapse starts from) is:

$$e_\xi=\zeta\frac{dmc_H^2}{dt_i\varepsilon_H}\left(1-\sqrt{1-\frac{(c_H-i_H)^2}{c_H^2}}\right);$$

and the missing drive, the quantum impulse is:

$$\lim\Delta_{quantum}=(1-\zeta)\frac{dmc^2}{dt_i\varepsilon_e}\left(1-\sqrt{1-\frac{(c-i)^2}{c^2}}\right)=0 \qquad \text{3G2}$$

The *Hydrogen* process has no cycles and the intensity of its electron process is of infinite low value. The question therefore is: how in this case its electron process drive is reaching its end status in order to start as neutron process?

The answer is that the electron process, with its infinite low intensity is the *drive* and the *end status* itself, incorporating both: the neutron process to be driven as the end product of the electron process and the electron process as the drive as well.
The electron/neutron process of the *Hydrogen* process,
(1) as electron process surplus results in conflict;
(2) as neutron process, is subject to be driven by other elementary processes.

The *Oxygen* process has its electron process surplus and the *Hydrogen* process also has its electron/neutron process surplus. The difference is that the *Hydrogen* process in fact never drives their neutrons and the *Hydrogen* cycles remain without increase. The only source for the generation of the *Hydrogen* process is the elementary evolution.
The *Hydrogen* process has its proton process intensity potential and "free" electron/neutron process to be driven. With the drive the number of the cycles of the *Oxygen* processes is increasing.
Each electron/neutron process of the *Hydrogen* process, driven by the *Oxygen* process results in a new *Oxygen* process, ready for driving other electron/neutron processes of the *Hydrogen* process again and again.
The number of the electron/neutron processes within the *Hydrogen* process has been used and the relation of the intensities of the proton and the neutron processes of the *Oxygen* process with the increase of the cycles is streaming to approach $\varepsilon_e^O = 1$ as well.
Diagram 3.4 shows the very efficient communication of the two elementary processes. The *Hydrogen* process is capable for the communication with high number of *Oxygen* processes.
Water, as the subject of the communication of the *Oxygen* and the *Hydrogen* processes results in the increased utilisation of the electron process surplus of the *Oxygen* process.

Oxygen cycles n_p – the number of the proton processes means the number of the cycles	*Hydrogen* process without cycle	There is a remaining proton process cover potential within the *Hydrogen* process and a limited electron process surplus within the *Oxygen* process and well. The last cycle of the *Oxygen* process is the only one, generating the remaining surplus – reason of the liquid status of the water.
$n_p^O + \Delta n_p^O$ n_n^O	n_p^H n_n^H	
$n_p^O + \Delta n_p^O + \Delta 1 n_p^O$ $n_n^O + \Delta 1 n_n^O$	n_p^H $n_n^H - \Delta 1 n_n^H$	
$n_p^O + \Delta n_p^O + \Delta 1 n_p^O + \ldots + \Delta N n_p^O$ $n_n^O + \Delta n_n^O + \Delta 1 n_n^O + \ldots$	n_p^H $n_n^H - \Delta N n_n^H$	

Diag. 3.4

Diagram 3.4

Water is in liquid aggregate status, because the speed of the quantum communication of the *Oxygen* process is less than the quantum speed value on the *Earth* surface. This is the reason water cannot be in conflict with the components of the *Air*, in gaseous status.

Water contains the accumulated proton process capacity of the *Hydrogen* processes, reserve of energy.

As the elementary quantum speed of the *Oxygen* process is less than the quantum speed on the *Earth* surface, water is extinguishing fire, in communication with it. Fire is a process of increased electron process conflict at the quantum speed of light on the surface of the *Earth*. Adding water on fire, water takes away part of the conflict or the conflict in whole, as its quantum drive is of less intensity and its quantum speed is of less value. As result of the communication, water is taking off intensity from the fire and this way is reducing the intensity of the conflict (fire). In the case the volume of the water is not sufficient, the communication goes different way and the fire is increasing the intensity of the water cycles and the water is steaming away.

3.4.2. S. 3.4.2

Steam is generated from water

$$\frac{dmc_w^2}{\varepsilon_w} = \frac{dmc_O^2}{\varepsilon_O} + \frac{dmc_H^2}{\varepsilon_H} \qquad \lim = 0$$ 3G3

As the elementary quantum drive is of infinite length in our space-time, the number of cycles are limited, since while the new cycle will obviously be an *Oxygen* one, it needs 10^{27-29} years in our space-time to happen.

Therefore for the assessment of the parameters, we have to calculate only with the processes of the actual communication.

In accordance with this:

$IQ_w < IQ_O$ and very likely $c_w < c_O$ and $\varepsilon_w > 1$, as $\lim \varepsilon_H = \propto$ and also $\varepsilon_O > 1$. 3G4

If water is heated up, it means:

$$IQ_{steam} = \frac{dmc_w^2}{\varepsilon_w \sqrt{1 - \frac{v^2}{c_w^2}}};$$

causing increased internal conflict and resulting in steam (gaseous) aggregate state. $IQ_{steam} > IQ_w$ 3G5

Without free space to expand, or in the case the instrument, tank, reservoir, container... remains of the same, limited or less volume, the intensity increase results in the increase (in conventional terms) of the pressure.

In order to withstand this increased pressure, or the conflicting impact of the increased *IQ* value of the steam, the container, holding the water/steam transformation shall be of elementary structure of increased internal intensity. The value of the quantum drive of the container shall be higher than the that of the generating steam:

$$IQ_{container} > IQ_{steam}$$ 3G6

With steam generation of unlimited growth, unlimited water source but limited space, there is no elementary process for the holder, which would withstand the unlimited electron process quantum impact of the generating conflict. (There is no "pure" elementary process with equilibrium elementary proton/neutron process status in the nature. Diamond is a mineral.)

If the container of the water is of a concrete structure, from selected mineral composition, giving water tight specifics for holding the water, the increased intensity of the steam process might destroy the water tightness of the construction.
Water tightness requires quasi equilibrium status of the concrete/mineral constructions. The equilibrium status reflects back external quantum impacts of the water.

Water has its conflicting (but limited) electron process quantum impact.
The quantum impact of the increased intensity of the generating steam might have its intensity capacity, impacting the elementary communication of the quasi equilibrium status. This depends only on the capacity of the energy source of the steam generation. A diamond-like concrete/mineral structure is the best to transform the water tightness to seam tightness.

S. 3.4.3

3.4.3

The principal basics of the behaviour of ***hydrocarbons*** are similar to those of the water; just the quantum speed and the intensity values of the quantum drive of the *Carbon* process are higher than those of the *Oxygen* process.

The electron process surplus of the *Carbon* process will be driving the "free" electron/neutron processes of the *Hydrogen* process, the same way as the *Oxygen* process does, resulting in *Carbon* process generation with electron process surplus. The intensity of the generating surplus however is different than that in the case of the *Oxygen* process. The reason is that the quantum speed of the *Carbon* process is higher not just than the quantum speed of the *Oxygen* process, but also than the quantum speed on the *Earth* surface.
The difference in the quantum speed values of the *Oxygen* and the *Carbon* processes explains the difference in the communication of the *hydrocarbons* with fire.

> As the operating quantum speed of the *hydrocarbons* becomes higher than the quantum speed (the speed of light) on the *Earth* surface, the communication of *hydrocarbons* with fire feeds the conflict, makes it more intensive: *Hydrocarbons* are increasing the conflict and the increasing conflict is deepening the fire.

The question still may be raised: why the *Oxygen* and why the *Carbon* processes?
The answer is simple: because these are the ones in the line of the elementary evolution with the highest *IQ* drive, electron process surplus and intensities.

S. 3.5

3.5
The impact of parallel cycles

Parallel cycles mean conflict on the direct or on the anti-sides. The higher the periodic number is, the higher is the number of the parallel cycles; the higher is the intensity, with increased speed of quantum communication.

While ε_x, the intensity of the electron process is about the internal relation of the elementary cycle: $\frac{dmc_x^2}{dt_i\varepsilon_x}\left(1-\sqrt{1-\frac{(c_x-i_x)^2}{c_x^2}}\right)$; (3H1)

the increase of the numbers of the elementary cycles,
acting in parallel, increases the conflict: $n\frac{dmc_x^2}{dt_{,i}\varepsilon_x} = \frac{dmc_x^2}{dt_{,i}\varepsilon_x\sqrt{1-(v^2/c_x^2)}}$; 3H2

Parallel elementary cycles are not about high number of single running elementary processes, rather parallel processes of the standard elementary process with common anti-electron process platform. Each of the parallel cycles generates quantum impulse.

Diagram 3.5 below demonstrates the necessity of the parallel cycles.
The *proton-electron-neutron-antineutron-antielectron-antiproton* periodicity cannot be managed within one and the same cycle. There is a need at least for *two cycles* in sequence for the continuity of the elementary process.

- the electron process cannot drive the neutron process within the same cycle;
- electron processes drive neutron processes in two cycles ahead of the actual drive.

cycle I	cycle II	cycle III	cycle IV	cycle V	cycle VI	cycle VII
p -1	p 1	p 2	p 3	p 4	p 5	p 6
e 1	e 2	e 3	e 4	e 5	e 6	e -1
n -2	**n -1**	**n 1**	**n 2**	**n 3**	**n 4**	**n 5**
an -1	an 1	an 2	an 3	an 4	an 5	an 6
ae 1	ae 2	ae 3	ae 4	ae 5	ae 6	ae -1
ap -2	ap -1	ap 1	ap 2	ap 3	ap 4	ap 5
p -1	p 1	p 2	p 3	p 4	p 5	p 6
to **III**	to **IV**	to **V**	**to VI**	to **VII**	to **I**	to **II**

- the *proton* and the *anti-neutron* processes run in parallel;
- the *electron* and *anti-electron* processes run in parallel;
- the cycles in fact start from the *neutron* and the *anti-proton* processes;
- the *anti-proton* process also marked by the same step number, but this is just for signalling that this process is combined with the *neutron* process;
- the *proton* processes on the bottom of each process cycle are the same *proton* process, just by 2 cycles later:
  I to III; II to IV; III to V; IV to VI; VI to VI+2; VII to VII+2
- there should at least 2 cycles to be for managing an elementary process;
- this is important to remember: the cycles represent continuous change.

Diagram 3.5 Diag. 3.5

The cycles represent the elementary processes in continuity without end. The end of a certain proton process means the immediate start of the next step, the electron process. At the moment of the expiration of the intensity potential of the electron process drive the collapse starts. The same way on the anti-direction.

The surpluses on both sides of the elementary cycles have their functions and do not influence the continuity of the elementary cycles. Therefore there will always be proton and anti-neutron covers for the neutron and anti-proton collapses whatever the intensity differences between the proton and the neutron processes, the anti-neutron and the anti-proton processes are.

The intensity difference between elementary phases may suggest the accumulation of processes within the elementary cycle; either proton or neutron processes. But this is not the case. Neutron processes cannot be left without drive and there is no accumulation of the anti-proton processes, "waiting" for to be driven as well.

Anti-neutron processes are the energy reserves of the *mechanical impact of gravitation*; proton processes are the energy *capacities of the quantum communication.*
Higher intensity means the process happens for less time than other processes.

- electron processes either end up as neutron process drives (independently what the original elementary process of the drive is) or represent *blue shift* quantum impact, generating conflict, until their intensity expires;
- anti-electron processes drive anti-proton processes, generate surplus, formulate the quantum membrane, establishing by that the intensity of the electron process on the direct side and do expire as the quantum impact of *gravitation.*

The duration of the process is: $\Delta t = \frac{1}{\varepsilon}$;

313 the direct cycle is: $\Delta t_p + \Delta t_n$; the anti-directions is: $\Delta t_{n-} + \Delta t_{p-}$;

In the case of proton process dominance the time balance of the two directions is:

314 $$\frac{1}{\varepsilon_p} + \underbrace{\left[\frac{1}{\varepsilon_p} + \left(\frac{\varepsilon_p}{\varepsilon_n} - 1\right)\right]}_{\text{neutron process}} = \underbrace{\left[\frac{1}{\varepsilon_p} + \left(\frac{\varepsilon_p}{\varepsilon_n} - 1\right)\right]}_{\text{anti-neutron process}} + \frac{1}{\varepsilon_p}$$

In the case of neutron process dominance the time balance is:

315 $$\frac{1}{\varepsilon_n} + \underbrace{\left[\frac{1}{\varepsilon_n} + \left(\frac{\varepsilon_n}{\varepsilon_p} - 1\right)\right]}_{\text{proton process}} = \underbrace{\left[\frac{1}{\varepsilon_n} + \left(\frac{\varepsilon_n}{\varepsilon_p} - 1\right)\right]}_{\text{anti-proton process}} + \frac{1}{\varepsilon_n}$$

The result in both cases is: $\Delta t_p + \Delta t_n = \Delta t_{n-} + \Delta t_{p-}$

- in other words, the anti-processes always control the elementary cycle:
  - the anti-neutron process exactly repeat the neutron process, just in the opposite direction;
  - the proton process is taken back by the anti-proton process and the cycle can continue in line with the standards of the elementary process;

The neutron process is the turning point. Why?
For interaction and communication, the neutron process is the one available or ready for elementary impacts. And the neutron processes are and can be driven by the infinite variety of the intensities of the elementary processes of the Periodic Table.

In the case of elementary communication the *IQ* drive is the one, which specifies the intensity of the electron process: the electron process with higher intensity of the impact has the priority.

The anti-neutron processes are consequences; the anti-electron processes are with controlling function: establishing the intensity of the electron process at one side and the intensity of the collapse of the anti-proton process on the other, in harmony with the quantum speed value and the standards of the elementary process. This way, the proton process is a consequence.

The quantum impact of *gravitation* is the surplus of the anti-electron processes.
There is a complexity of 7 cycles within the Diagram 3.2. There are infinite numbers of cycles in the nature in each case of the elementary processes. The change and the cycles are continuous. The most important message of Table 3.1, Diagram 3.2 with 7 cycles and Diagram 3.3 with 21 cycles is that the continuity and the normal operation of the elementary processes requires at least <u>*two*</u> parallel running cycles. This corresponds to the *Helium* elementary process with PN 2, the last process with complete structure.

Ref. Diag. 3.2 Table 3.1 Diag. 3.3

The *plasma* and the *Hydrogen* processes belong to each other. They both represent the half of a normal elementary cycle. The *Hydrogen* process provides the proton and the electron/neutron processes to the inflexion; the *plasma* is the neutron/anti-neutron inflexion itself and ensures the anti-electron and anti-proton processes of the change.

3.6
The net of elementary processes

S. 3.6

Minerals are the combination of elementary processes in infinite numbers and in variety for finding the best and optimal balancing structure of the communication at different levels of the cooling process of the plasma.

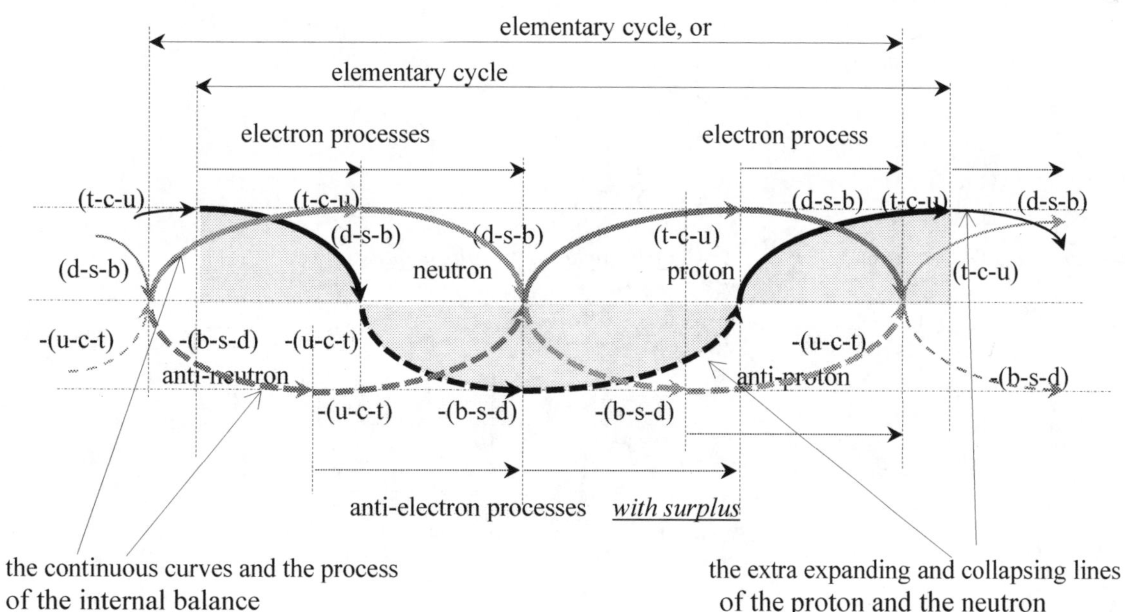

Diag. 3.6

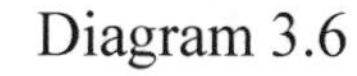
Diagram 3.6

For melting certain "element" out of a mineral, there is a need for the generation of electron process conflict within the mineral. The conflict is impacting the elementary process and makes the element liquid, as this is the product of the conflict.

The conflict of the electron process means heat generation. The conflict gives over the conflicting *blue shift* impact to the cooling process – the conflict is expanding and disappears. The melted out elementary process appears in its clean status.
The clean elementary status is also a certain net of elementary processes. The proton process has an extra expanding *t-c-u* quark chain; the neutron process has an extra *d-s-b* quarks line. The two quark processes, as *expansion-collapse* cycle are in balance with each other, represent the same space-time within the "clean" elementary process.

In the case of a homogeneous elementary structure of a "clean elementary process", the extra balance is driven by the electron process of the elementary process itself. The elementary process has the equal and the same quantum speed value.

There is a neutron process dominant elementary process pictured on the Diagram 3.7.
(The intensity of the collapse is higher than the intensity of the expansion.)
The anti-neutron process generates anti-electron process surplus.

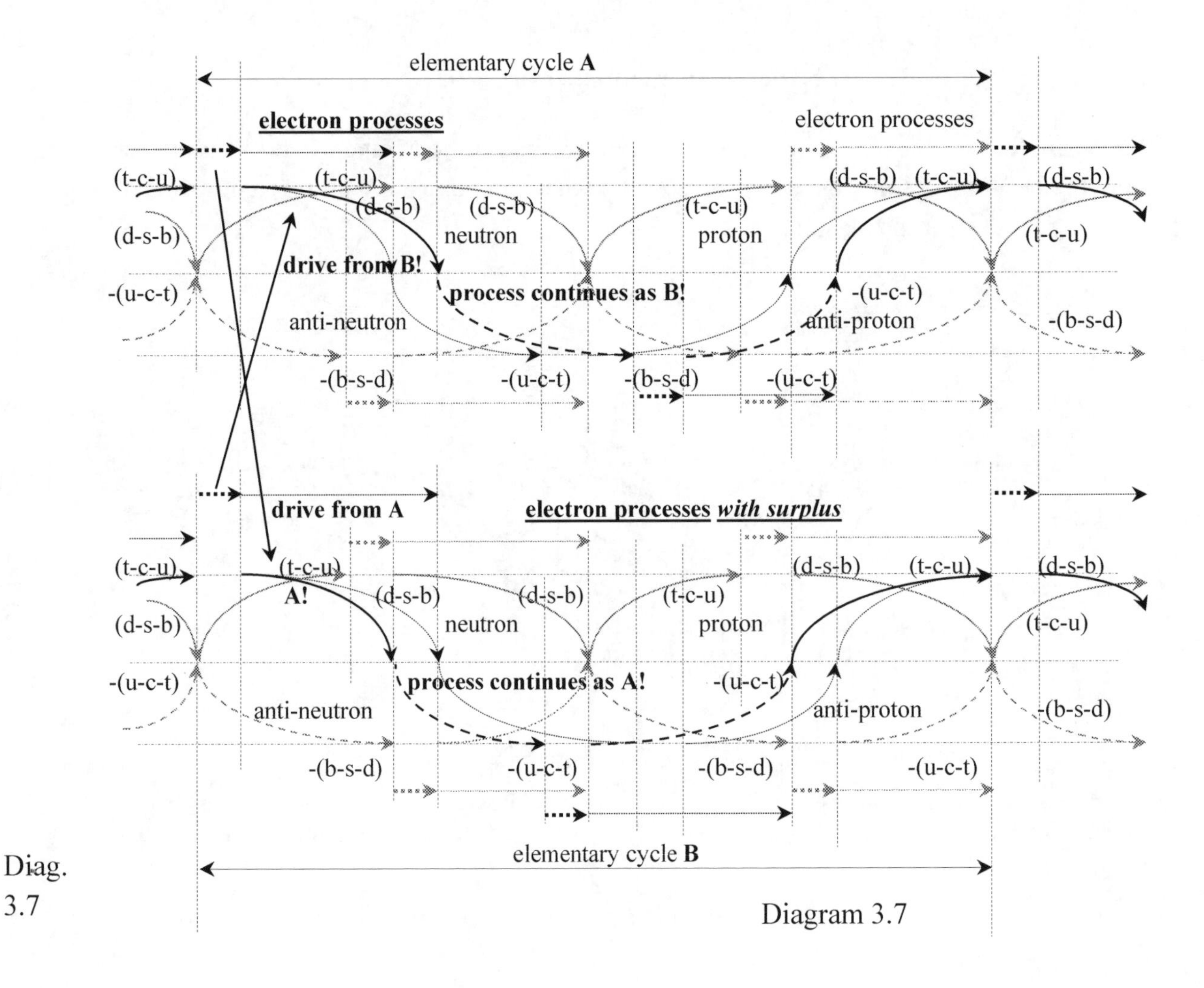

Diag. 3.7

Diagram 3.7

In the case of elementary communication, as Diagram 3.7 above demonstrates it, the extra balancing lines drive the neutron process of the other elementary process. The drive is changing in each cycle, as the diagram demonstrates it. The electron process of A takes over the drive of B as the intensity of the electron process of A is higher. Once the communication starts, the neutron process of A becomes in drive demand. B provides it.

The extra lines become replaced. B will be acting in A and A will be acting in B.
At the end of the cycle, therefore the extra line in B results in neutron process dominance, corresponding to A. The extra line in A represents proton process dominance. (The anti-electron processes are controlling and re-establish the process.)
The impact of the higher intensity will be acting: the A becomes A, and the B becomes B again. And so on.

Different quantum speed values are acting in one and the same elementary composition. This makes minerals and compositions unique. The reason of oxides, silicates and others is the tendency of the nature to create balanced elementary structures.
The acting drives depend on the intensity and also on the number of the parallel cycles.

The communication of the *Magnesium* for example, with its 12 parallel cycles and the *Oxygen* with its 8 cycles establish the specific relation of extra lines, as 8 is communicating with 12.
The principle is the same, but the result is different in the case of the communication of the *Oxygen* process with the *Uranium* process with 92 cycles.

Matter is the matrix of quantum conflicts, the *net* of infinite number of elementary processes with the infinite variety of the quantum speed of communication.

Quantum communication is about the communication of various elementary compositions: (1) only of neutron process dominant elementary process, like *NaCl*, the *Sodium Chloride* and others; (2) only proton process dominant elementary processes like *water*, *smoke* and others; (3) elementary processes with both, neutron and proton process dominance.

Classical elementary communication means the communication of the two opposite sides. This type of communication establishes and keeps the process in natural equilibrium status, the genuine objective of all elementary communications. The electron processes drive each other's neutron processes and the anti-processes re-establish the elementary process.

The electron and the anti-electron processes are impacting the neutron and the anti-proton processes of other elementary process. The equilibrium will be kept on, during the continuity of the communication, as the reciprocal impact will be acting.
This is in fact communication of space-times.

4
Elementary communication in practice
The benefit of the quantum impact

Ref. Diag. 3.7

The *granite* is a unique mineral. Its integrated elementary operation is close to equilibrium. The best reference to its elementary structure is Diagram 3.7, where almost all of those elementary processes of the diagram are part of the elementary composition of the *granite*. The cross communication between these elementary processes helps to reach the balanced status at integrated level. *Granite* is the building stone of pyramids.

Let us to examine, what the elementary composition of the *granite* mineral is and what the internal quantum communication of this structure is about.

Ref. Book 5. S.8. Table 6.1 Table 4.1

The pyramid has been built up from *granite* minerals with general formula of the *Feldspars* with the main components of: $KAlSi_3O_8$ – $NaAlSi_3O_8$– $CaAl_2Si_2O_8$

Component	%
SiO_2	- 72.4 %
Al_2O_3	- 14.42 %
K_2O	- 4.12 %
Na_2O	- 3.69 %
CaO	- 1.82 %
FeO	- 1.68 %
Fe_2O	- 1.22 %
MgO	- 0.71 %
TiO_2	- 0.30 %
P_2O_5	- 0.12 %
MnO	- 0.05 %

Table 4.1

Elementary processes	ε_x	c_x	$IQ = \frac{c_x^2}{\varepsilon_x}$
Oxygen	1.01533	*299711**	8.847E+10
Sodium (Na)	0.93084	313018	1.053E+11
Magnesium	0.98990	303536	9.307E+10
Aluminium	0.94697	310340	1.017E+11
Silicon	1.00898	*300652*	8.959E+10
Phosphorus	0.95283	309348	1.004E+11
Potassium (K)	0.95932	308335	9.910E+10
Calcium	1.01112	*300334*	8.921E+10
Titanium	0.86133	325403	1.229E+11
Manganese	0.84667	310340	1.138E+11
Iron (Fe)	0.88347	321300	1.169E+11

Reference: "The Quantum Impulse and the Space-Time Matrix" 2015

All elementary processes in Table 4.1 above – except *Ti, Mn* and *Fe* – (either are of proton or neutron process dominance), are close to the equilibrium status, with quantum speed values above and below 300,000 km/sec, with intensities close to each other. The elementary communication of the components gives solid, almost unbreakable structure to the *granite* mineral.

The dominant *Silicon-dioxide* content means the utilisation of the surplus of the electron process drive of the *Silicon* process with the increase of the electron process surplus of the *Oxygen* process. At the same time, the wide elementary communication within the *granite* structure excludes the *Oxygen* process dominance and the electron process surplus of the *Oxygen* process.

Let us to examine as example, what the cross-communication resulting in *Titanium-dioxide* is about?

The "invitation" – during the growth of the *granite* mineral under the intensity impact of the *plasma* – is coming from the electron process *blue shift* conflict, generated by the low intensity electron process drive of the *Oxygen* process. When the *Titanium* drive is acting, the continuity of the elementary processes makes the low intensity *Oxygen* drive also acting.
Neutron processes only exist if they are driven. Neutrons cannot and do not "wait" for the drive of the collapse to come. The proton process cover, the electron process drive and the collapse are acting in parallel. Once the electron processes of the *Titanium* with dominant intensity, drive the neutrons available, within both, the *Titanium* and the *Oxygen* elementary processes, the electron processes of the *Oxygen* have the chance (even with less intensity value) to drive the "remaining" neutrons (without drive), including the ones within the *Titanium* process as well.

The *Titanium* process has 22 acting in parallel cycles within its elementary structure.
The *Oxygen* process has only 8. The *Oxygen* process is with electron process surplus of low intensity impact. In the communication with the *Titanium* process, even the generating neutron processes of the *Oxygen* process will be driven, as priority, by the electrons of the *Titanium* as they have the controlling intensity dominance within the communication.
But, with reference to Table 3.1 of the previous section, the gradient of the loss on the intensity of the electron process of the *Titanium* is significant, while those of the *Oxygen* process keep their quasi stable (but low value) intensity status. The *Oxygen* process this way has the chance to drive the neutrons of the *Titanium* process as well. This is the reason of the cross-communication. Both elementary processes are repeating the impacts the same way again and again.

Ref. Table 3.1

The cross-communication of the two elementary processes is a natural need. The constant internal conflict is the consequence of the acting electron process surplus of the *Oxygen* process. This internal conflict is also the reason the *Titanium-dioxide* mineral is a powder.

The principle of the communication of all *oxides* of Table 4.1 on the previous page is similar. The only difference is the communication of the proton process dominant elementary processes, like *Calcium* and *Silicon*, where the acting surplus is turning into deficit again and again. This means permanent invitation for the *Oxygen* process to use its electron process surplus.

Ref. Table 4.1

During the natural formulation of the *granite* mineral, under the constant quantum impact of the plasma and *gravitation*, the dominant process of the cross-communication is the *Silicon* process, with the highest value of the electron process surplus.

In communication with the *Al, K, Na, Fe, Mg, Ti, P* and *Mn* processes and with their oxides, the conflict, generated by the *plasma* impact and *gravitation*, plus the conflict of the *O*, *Ca* and *Si* processes results in all over elementary communication, with the central point of the *Silicon* process. This is presented in Table 4.2 on the next page.

Ref. Table 4.2

The *Silicon* process communicates in two directions:

(1) – *Oxygen* and *Calcium*; (2) – all other elementary processes.

	O	Na	Mg	Al	***Si***	P	K	*Ca*	Ti	Mn	Fe
PN	8	11	12	13	14	15	19	20	22	25	26
intensity	1.015	0.931	0.990	0.947	1.009	0.953	0.959	1.011	0.861	0.847	0.883
surplus	+	-	-	-	+	-	-	+	-	-	-
*IQ**E+10	8.8	10.5	9.3	10.2	8.96	10.0	9.9	8.92	12.3	11.4	11.7
part		3.6%	0.7%	12.4%	72.4%	0.1%	4.2%	1.8%	0.3	0.05%	2.8%

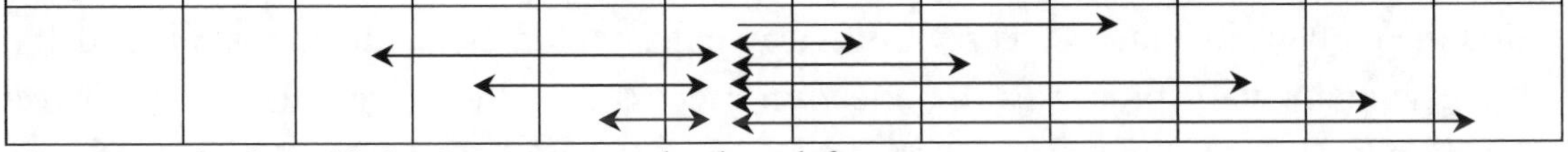

Table 4.2

The communication with the *Oxygen* process has been left out.

Table 4.2

The *Silicon* process is inviting the electron process drives of all elementary processes and as response, provides the missing drive deficit, consequence of the communication. The *Calcium* and the *Oxygen* processes have similar communication strategy, but the intensity of their quantum drives are of less value.

The all over cross-communication is controlled by the general principle to formulate elementary composition, the closest as possible to equilibrium. The *Oxygen* process is the elementary component in the widest use. The largest surplus in electron process drives belongs to the *Silicon* process. The *Titanium* process has the highest quantum drive intensity, and the *Aluminium* process is the closest to the equilibrium state among the neutron process dominant elementary processes.

S. 4.1

4.1
The example of the *communication*: concrete structures

Mixing limestone $CaCO_3$ and silica SiO_2 with water, the conflict increases the intensity of the communication. There are also other minerals and elementary processes in the cement, as Al_2O_3, Fe_2O_3, MgO and SO_3 but for simplicity we limit now our assessment to the *Calcium* and the *Silicon* components only. (The *Calcium* process in the limestone and the *Silicon* process in the silica will be representing the minerals themselves, as they are the ones with the highest quantum speed values within the mineral and have the dominance in the communication.)

The conflict means, the *Ca* and the *Si* elementary processes with higher quantum speed and intensity values try to use the neutron processes of the *Oxygen* of the water. The *Si* process also has its impact on the neutrons of the *Ca* process as well.

Because of the higher *IQ* drives of the *Si* and the *Ca* processes the intensity of the *Oxygen* processes has relative slowdown within the mix. Because of the less than available utilisation of the electron process potential of the *Oxygen* process, the developing intensity surplus results in conflict and the conflict generates heat.

By constant mixing the heat expires, but the composition might also need more water. The watering, if necessary, gives the chance to all *Ca, Si* and *O* components to further utilise the electron process surplus the best possible and optimal way.

Silicon		*Calcium*		*Water*				The *O* in the water has its full cycle with electron process surplus and the *H* is with its constant proton process surplus.
				Oxygen		*Hydrogen*		
e	p	e	p	e	p		p	
	n_{Si}		n_{Ca}		n_O			$IQ_{Si} > IQ_{Ca} > IQ_O$ driving neutrons in line with the intensity potential, resulting
$n_{Si}e$	$n_{Si}p$	$n_{Ca}e$	$n_{Ca}p$	n_Oe	n_Op		p	$n_{Si}e > n_{Ca}e > n_Oe$ number of the electron processes
$\Delta n_{eSi} = (1.00898-1)\, n_{Si}e$ $\Delta n_{eCa} = (1.01112-[1+x])\, n_{Ca}e > 0$ $\Delta n_{eO} = (1.01533-[1+0.01533])\, n_Oe = 0$								The neutrons of the *O* are driven by the *Si* and *Ca* processes; the electron processes are in surplus.
	$n_{Si}n$		$n_{Ca}n$		$(n_O + +\Delta n_O)$		$p + \Delta p$	Additional water for utilising the electron process surplus of the *Ca* and *Si* processes – drive the additional neutrons!
$nl_{Si}e$	$nl_{Si}p$	$nl_{Ca}e$	$nl_{Ca}p$	nl_Oe				Electron processes remain in increased surplus.
$\Delta nl_{eSi} = (1.00898-1)\, n_{Si}e$ $\Delta nl_{eCa} = (1.01112-[1+0.1112])\, n_{Ca}e \cong 0$ $\Delta 1n_{eO} = (1.01533-[1+0.01533])\, n_Oe = 0$								The neutron processes of the *O* and partially of the *Ca* have been utilised by the electron process surplus of the *Ca* and partially by the *Si* processes.

Diagram 4.3

Diag. 4.3

The neutrons are driven as the acting IQ controls it.

$c_{Si} = 300{,}652$ km/sec; $IQ_{Si} = 8.959E+10$; $\varepsilon_{Si} = 1.00898$

$c_{Ca} = 300{,}334$ km/sec; $IQ_{Ca} = 8.921E+10$; $\varepsilon_{Ca} = 1.01112$

$c_O = 299{,}711$ km/sec; $IQ_O = 8.847E+10$; $\varepsilon_O = 1.01533$

The electron(s)/neutron(s) of the *Hydrogen* process, composing water have already been driven by the *Oxygen* process. But the *Oxygen* process still has its electron process surplus in water. (Each driven electron/neutron processes of the *Hydrogen* process generates *Oxygen* process and the *Oxygen* process is with electron process surplus.)

The *Si* and the *Ca* processes drive the neutrons of the *Oxygen* process within the water again and again. This way the *Oxygen* process finds itself in permanent electron process conflict. This conflict is the reason of the heat generation. The conflict and the increased quantum speed of the environment makes it partially gaseous. The *Hydrogen* process, or better to say, the proton processes of the *Hydrogen* process, with infinite low quantum speed value, remains within the concrete mix as the energy potential of the structure.

The utilisation of the electron processes of the *Si* and *Ca* processes for driving the neutrons of the *Oxygen* process reduces the conflict and cools down the mix. The mix becomes hardened and cross-combined. Space-times of the elementary processes have become tight in communication.

The principle of the communication is easy.
The stronger one in the intensity of the electron process quantum drive, drives the neutrons. At the start of the communication all elementary components, *Ca, Si, O* (water) are with electron process surplus. All proton process communication follows the line of the drive. The neutron processes are neutral. Watering increases the cross communication. The *Silicon* process is utilising its increased electron process quantum drive as much as possible in each cycle. The *Calcium* process has its partial use, the *Oxygen* process, with less intensity remains in surplus.
The cycles repeat the communication.

With sufficient watering the *Oxygen* process either will be balanced or becomes gaseous; the *Calcium* process uses its electron process surplus in full; the last cycle of the *Silicon* process remains with not used electron process surplus.
The *Hydrogen* process provides a kind of energy intensity by "managing" its proton process impacts of infinite low quantum speed in the mix.

Adding elementary processes with neutron process dominance into the mix, like *Cu, Ti, Fe* does not just increase the intensity of the communication, but improves the balance of the formulating matrix as well. The *Cu, Ti* and *Fe* processes have increased quantum drives; *IQ* values from $IQ_{Fe} = 1.169E+11$ to $IQ_{Cu} = 1.258E+11$.

*

The two different types of the elementary structures (proton and neutron process dominance) is also reason of the *conflict*:

- the conflict is initiated by the *Si-Ca-O(water)* elementary composition, *with proton process dominance,* denoted as ***A***;
- the conflict means the speed of quantum communication of the components becomes increased;
- speeding up means breaking the internal balance of the elementary processes;
- the anti-processes immediately "feel" the decline from the standard values, which must be restored;
- the goal of the elementary communication is ***resolving/eliminate*** the conflict.

The *IQ* value of each of the *Cu, Ti,* and *Fe* processes with *neutron process dominance* is higher. These components are denoted as ***B***,

- therefore ***B*** will be driving the neutrons of the elementary processes of ***A***;
- at the same time its standard neutron process remains without sufficient drive.

This increases the conflict within ***A***, otherwise with low electron process intensity, because

- the external drive increases the electron process surplus, which increases the intensity of the anti-electron process of ***A***;
- the intensity of the anti-electron process of ***A*** however cannot be increased, as the anti-proton process has its constant intensity value;
- therefore ***A*** limits the external impact: the *IQ* drive of ***A*** drives the neutrons of ***B***, having been left without sufficient drive.

  In normal circumstances, without the mix, the intensity of the *IQ* drive of ***A*** would be not capable to drive the neutrons of ***B***.

B:	**A:**
$\frac{c_B^2}{\varepsilon_B} + \frac{c_x^2}{\varepsilon_x} = \frac{c_{B1}^2}{\varepsilon_B}; \quad c_{B1} > c_B$ The increased quantum speed is result of the conflict, but having $\varepsilon_B = const$ - as being controlled by the anti-electron process - the electron process drives neutrons, available in the communication. [The conflict is not at the level, which would result in the modification of the elementary process.]	The elementary composition has electron process surplus but the quantum drive is of less intensity. Therefore the neutron processes can also be driven by an increased drive from aside. (4A1) The external impact of ***B*** drives certain number of neutrons of ***A***. This increases the conflict within ***A*** as well, since the surplus here also becomes increased: $c_{A1} > c_A$ The increased surplus initiates the increase of the intensity of the anti-electron process as well. But it should be constant otherwise the elementary process would be destroyed (in the case of fire as example). The anti-electron processes of ***A*** control the case and limit the external drive from ***B***, the way
The extra use of the drives, driving the neutrons of ***A*** increases the number of the elementary cycles, the increased quantum speed is working out to be normal again.	the electron processes of ***A*** drive the neutrons of ***B*** left in fact without drive. — $n_B \frac{c_B^2}{\varepsilon_B} = n_A \frac{c_A^2}{\varepsilon_A}$ (4A2) and as $\frac{c_B^2}{\varepsilon_B} > \frac{c_A^2}{\varepsilon_A}$; it would require: $n_A > n_B$ (4A3)
The anti-electron processes control the processes of both types and keep the *IQ* drive of the anti-proton processes at standard and equal values.	But ***A*** can drive at maximum n_B, the consequence of the increase of the surplus.

The communication will be in a dynamic balance, responding to the demands of the permanent increase-restore impacts of both sides. Once the conflict is over, the relation of ***A*** and ***B*** corresponds not just to the latest balance status, but the communication of ***A*** and ***B*** tries to establish and keep the overall balance, acting.

This is a strong binding force of the two.

A+B in balance together composes a strong solidified elementary structure.

A and ***B*** in contact establishes certain quantum communication relation, which exactly corresponds to their elementary characteristics, but different than their separate operation:

- ***A*** has its own standard ε_A intensity and c_A quantum speed values with less electron process surplus within the elementary cycle;
- ***B*** also has its own standard ε_B intensity and c_B quantum speed values with less anti-electron process surplus.

The two elementary processes together have a more balanced elementary structure!

With the less remaining electron process surplus of the cross communication of ***A*** and ***B*** results in a concrete structure with increased hardness and water tightness – a closer to balanced status elementary composition.

As the above example proves, the communication is a natural need for both sides, for the proton and also for the neutron process dominant elementary processes.

- The communication starts once the conflict is initiated by the water!
- The neutrons are the ones to be driven.
- There is no difference where the elementary drive is coming from.
  In proton process dominance the drive is of low intensity, therefore there is not just electron process surplus is available in this case within the elementary process, but the neutron processes can be driven by a higher driving potential from other elementary process.
  In the case of neutron process dominance the electron and the proton processes are fully used even without elementary communication. The communication however increases the option for their electron process drives.
- Communication happens only in the case of conflict. Elementary processes otherwise would not be communicating. Elementary processes would keep their standard structure. The proton process dominant elementary processes are the ones generating the conflict:
  - if the natural conflict is not sufficient for initiating the communication, it can be increased by external impact – by *heating*;
  - heating is additional electron process quantum impact, therefore it increases the conflict, as it speeds up the quantum communication;
- With more neutrons driven, the number of the elementary cycles is increasing.
- The portion of the anti-electron process surplus will be remaining as it is necessary for the standards of the elementary process, but the communication results in better utilisation and higher efficiency.

This is not about a full balance and far not about the equality of the intensities of the proton and the neutron processes (as it is within the diamond structure), rather it is the case when the conflicting electron process surplus, otherwise accumulating on the direct side becomes utilised.

The conflicting electron process surplus does not weaken the concrete structure, but:
if the elementary processes within the concrete structure with increased quantum speed value (*Si*, *Ca*) could not work out their electron process surplus (because of the lack of

the watering, which otherwise brings the neutrons of the *Oxygen* process in, to be driven) – the structure becomes of less stability.

The neutrons of the *Oxygen* process are driven and each driven *O* neutron generates either *Ca* or *Si* process with electron process surplus.

The absolute volume of the concrete structure is important.

If a certain volume consist n_{Ca} and n_{Si} number of elementary cycles, corresponding altogether to mass *m*, with additional neutron processes (with the watering) the number of the completed cycles is increasing, while the relative content of the electron process surplus within the acting *Si* and the *Ca* processes remains the same.

Without sufficient watering, the surplus of the *Si* and the *Ca* processes could not result in proper structuring: the existing internal conflict weakens the binding. The communication with water results in proper solid structure.

More water means higher number of completed *Si* and *Ca* processes, obviously always with remaining surplus. Too much water means too many acting processes, too many conflicts – the mixture becomes liquid. Less water results in less stability or cracks. Once the solidification has been completed, additional water does not communicate with the mix, but may have its impact.

- In the case while the water content of the mix is sufficient, but the solidified concrete structure is only with $CaCO_3$, SiO_2, Al_2O_3, Fe_2O_3, MgO and SO_3Ca minerals, there is a remaining acting electron process surplus in the structure. The composition is not in full balance and it becomes not water tight.
- Preparing the mix with *Ti, Cu* and other elementary components with increased neutron process dominance, the electron process surplus becomes utilised and the composition becomes balanced. The concrete construction is water tight.

The key point of any elementary communication is the *conflict*.

The intensities of the communicating elementary processes are increasing in the conflict. The numbers of the generating elementary processes are growing, while the anti-processes keep the elementary standards.

4.2 Temperature means the *conflict* of the electron process

S. 4.2

The electron process surplus of the *water*, the *plasma* and *gravitation* are the main initiators of the elementary communication.

- The water is with the electron process surplus of the *Oxygen* process and the proton process intensity potential of the *Hydrogen* process;
- The *plasma* is with the highest intensity of the anti-electron process *blue shift* impact;
- *Gravitation* is anti-electron process *blue shift* quantum impact.

The impact of watering is discussed in Section 3.4.1

Ref. S. 3.4.1

The electron process is the tool of the elementary communication and the subject of elementary conflicts.

Speaking about *temperature*, it seems the definition is obvious.
But what the temperature is about?
The temperature is no other than the physical parameter of the level of the <u>*conflict of the electron processes*</u>. Higher temperature means higher level of conflict, no conflict at the end means absolute zero temperature.
If the *plasma* status means electron process conflict of infinite high intensity, infinite high temperature and infinite short time, the *Hydrogen* process should mean infinite low temperature and infinite long time. (In our circumstances the quantum speed of the communication of the *Hydrogen* process has been speeded up to the quantum speed of the *Earth* surface.)
The *Hydrogen* process / *plasma* inflexion is of infinite high intensity indeed.

4A4 The conflict means: $n\frac{dmc_x^2}{dt_i\varepsilon_x}\left(1-\sqrt{1-\frac{(c_x-i_x)^2}{c_x^2}}\right)=\frac{dmc_x^2}{dt_i\varepsilon_x\sqrt{1-\frac{v^2}{c_x^2}}}\left(1-\sqrt{1-\frac{(c_x-i_x)^2}{c_x^2}}\right)$;

4A5 where $n=\frac{1}{\sqrt{1-\frac{v^2}{c_x^2}}}$; while the values of the intensity coefficients of the elementary processes of the conflict remain as they were: $\varepsilon_x=\frac{\varepsilon_p}{\varepsilon_n}\sqrt{1-\frac{(c_x-i_x)^2}{c_x^2}}$

The increase of the temperature in this case means a kind the generation of virtual electron process surplus, while the intensity coefficient of the elementary process(es) remain(s) not changed. It results in the change of the aggregate status of the processes to liquid or gaseous, depending on the level of the conflict.

S. 4.2.1

4.2.1. <u>*Does the intensity increase modify the time flow?*</u>

The answer is: it depends on the case.
With reference to the previous example, in the case of elementary processes, the conflict results only in the change of the aggregate status of the elementary process.
(The time flow and the space-time of elementary processes between the *plasma* and the *Hydrogen* processes are different anyway. And the infinite high intensity of the conflict of the electron and the anti-electron processes is the one resulting in the plasma status itself. The *plasma* has its own communicating impact with the "external environment", with those processes out of the plasma status, at the boundaries of the plasma. And this relation in general means not just decreasing temperature, but also speeding up time flow.)

If the conflict is about the acting quantum impacts of our quantum communication, about quantum signals, generated by our tools of communication, the consequence is different.

Each mobile phone, each computer, each email account, each instrument of the radio- and telecommunication of our modern life increases the conflict within our space-time.
The speed of the quantum communication in our space-time on the *Earth* surface is the speed of light in its conventional understanding. All our signals are generating at this

speed, all our signals generate conflict around us and all new signals deepen the conflict at this speed value. Each generating conflict increases the temperature.

Is this conflict and intensity increase slowing down the time flow on the *Earth* surface?
Do we have the chance with this intensity increase to slow down our life tempo?
Do we have the chance to fall behind the "normal time count" of the *Sun-Earth* relation, because of our artificially generated conflict of increased intensity on the *Earth* surface?
Does this conflict around us, slow down our time count and makes our life longer on the *Earth* surface?
The answer on all these listed options is: *no*! In the contrary.
We – as each of us is a unique personal composition of elementary processes – are not subjects to any intensity increase. In the contrary: additional internal efforts should be generated in order to withstand the increased external load, results of external conflicts. We feel only the heat, but the impact is more sophisticated and might be with damages.

Our environment is warming up with huge tempo. And the main reasons of the climate change are not the still operating power plants, rather the increasing number of all kinds of communication devices.

4.3
The quantum communication of the *pyramid*

S. 4.3

depends on the relations of the territories of the casing and the basic surfaces and other factors

The four sides of the pyramid generate electron process conflict within the structure of the pyramid, even with quasi-balanced *granite* mineral. The generating conflict is source of energy. The best for the utilisation of the conflict are actually minerals with balanced elementary structures, as in this case all generating energy can be used. The conflict is given off through the four sides of the pyramid, at different levels, in any direction.

There are x levels, with reference to Figure 4.1 in line with the height of the pyramid.
The absolute value of the loss (given off) at the first level is: $E_{n-1} = E_g - E_1 = (n_p - n_1)e_p$; 4B1

The electron process conflict, caused by the quantum impact of gravitation is given off not just by the horizontal surfaces of the pyramid, but also by the vertical ones as well.
The conflict is conflict, which propagates in any direction.

The intensity of the communication of the pyramid through the surfaces of its structure towards the external environment corresponds to the standard intensity of the quantum impact of gravitation, acting in its normal way.

The intensity of the quantum impact of gravitation is: $e_g = \frac{dmc_p^2}{dt_i \varepsilon_p}\left(1 - \sqrt{1 - \frac{(c_p - i_p)^2}{c_p^2}}\right)$; 4B2

This intensity impact is going through the mineral structure of the pyramid.

The gravitation impact in its full value through the surface of the base is:

4B3 $$\sum e_g = n_p \frac{dmc_p^2}{dt_i \varepsilon_p}\left(1-\sqrt{1-\frac{(c_p-i_p)^2}{c_p^2}}\right) = n_p \cdot e_p;\ n_p \text{ is proportional to the basic territory.}$$

for providing the standard quantum impact of gravitation at the periphery, the supply form the *Earth* shall be *1.6366* times more

r

a

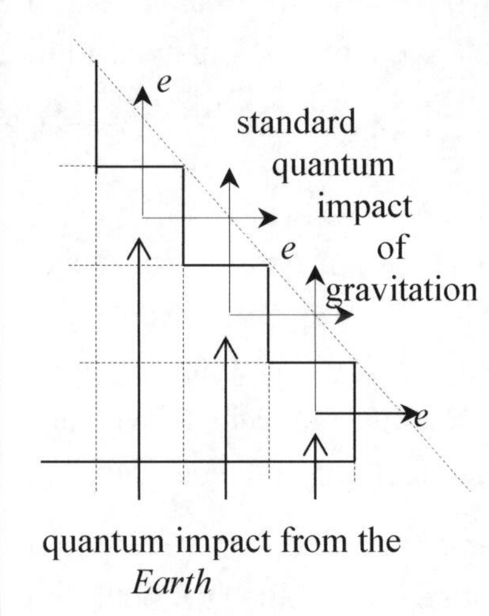

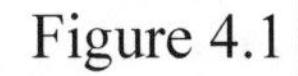
Fig. 4.1 Figure 4.1

e_{cas}

r

a

e_{basic}

The *approaching* impact of *gravitation* is the same at each points of the *pyramid.*

The relation of the territories of *the casing* and *the base* is equal to:

4B4 $$\frac{S_{cas}}{S_{basis}} = \varphi = 1.6189; \quad \text{and} \quad \frac{a}{r} = 1.57;$$

4B5 $$4a = 2r\pi$$

The relation of the *half of the perimeter* of *the base* and the height is equal to π:

4B6 $$\frac{2a}{r} = \pi; \quad \text{and } a = \frac{\pi}{2}r; \quad \text{and } r = \frac{2a}{\pi};$$

The relation of the territories of the base and the summarised *four* sides is:

4B7 $$S_{cas} > S_{basis}$$

The absolute value of the quantum impacts through the base and the sides are equal:

4B8 $$E = E_{basis} = E_{cov}$$

Obviously, the intensity of the incoming and leaving intensity values are different:

4B9 $$e_{basis} = \frac{E}{S_{basis}}; \quad \text{and} \quad e_{cas} = \frac{E}{S_{cas}};$$

Fig. 4.2 Figure 4.2

This relation is different, as the real territory of the casing in the step by step format of all four sides is:

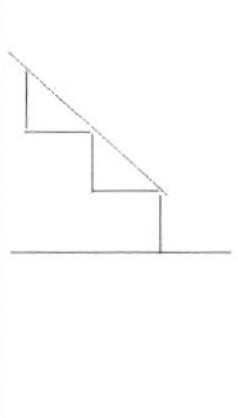

Fig.4.3

$$S_{cas} = a^2 + 4\frac{1}{2}\frac{a}{2}\frac{2a}{\pi} = a^2\left(1+\frac{2}{\pi}\right); \text{ or} \qquad \text{4B10}$$

$$S_{cas} = \frac{r^2\pi^2}{4} + 4\frac{1}{2}r\frac{r\pi}{2\cdot 2} = r^2\pi\left(\frac{\pi}{4}+\frac{1}{2}\right)=1.6366 \qquad \text{4B11}$$

Fig. 4.3

For having the conflicting *e* intensity impact at the periphery of the pyramid all around in vertical and horizontal directions, equal to the natural quantum impact of gravitation, the internal "supply" through the surface of the base from the *Earth* shall be more.

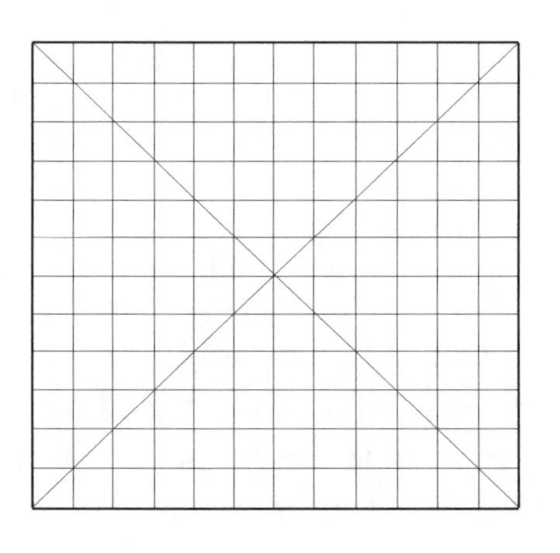

Fig. 4.4

The quantum system of the pyramid is transferring the quantum impact of gravitation from the base to the surface. Therefore each of the selected points (Fig.4.1) has its supplying impact through the base.

This impact shall be of increased value (more) as the surface of the cover is more. The increased volume results in increased quantum speed value, which means a *space-time* of increased intensity and quantum speed within the pyramid.

Fig. 4.4

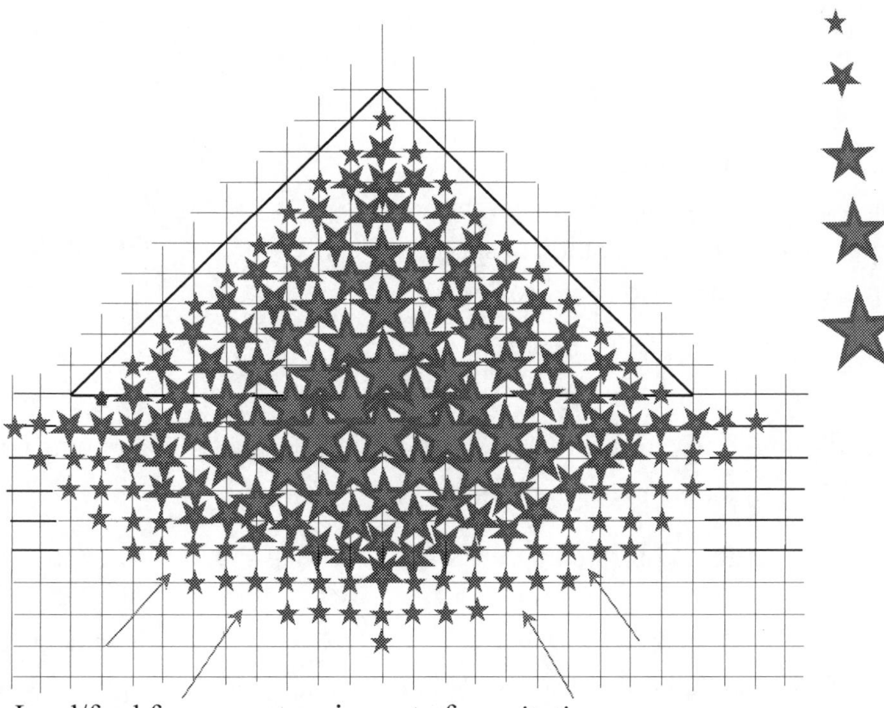

Load/feed from quantum impact of *gravitation*.

Figure 4.5

Stars mark the intensity of the quantum membrane, increasing towards the centre.

The intensity and the quantum speed values on the surface of the pyramid are quasi equal to the values on the *Earth* surface.

Fig. 4.5

[The closer the elementary balance of the stones of the pyramid structure to equilibrium is, the higher is the efficiency of the energy generation. In the case of minerals with proton or neutron process dominance, there is either an already existing conflict within the mineral, or it is less sensitive to any external conflicting impact. The impact from gravitation is the same, but the structure first of all controls its internal balance.]

Ref. QISM S.8 — The distribution of the electron process *blue shift* conflict within the pyramid is from Figure 8.5 of the book "*The Quantum Impulse and the Space-Time Matrix*".

The *IQ* drive of the *granite* mineral of the pyramid corresponds to:
$IQ_p = 9.064E{+}10\ km^2/\sec^2$, with $c_p = 301{,}699\ km/sec$ quantum speed and $\varepsilon_p = 1.00421$.

The data demonstrate how close the internal mineral structure of the pyramid to the equilibrium status is. The *IQ* drive of the quantum impact of gravitation is: 8.9875*E*+10.

4C1 — The relation of the two *IQ* drives is: $\dfrac{IQ_p}{IQ_E} = \dfrac{9.064}{8.987} = 1.00852$

All data are from the book "*The Quantum Impulse and Space-Time Matrix*" (2015)

Having the quantum impact equal to the intensity of the of natural impact of *gravitation* at each of the points of the pyramid surface and in any direction, the intensity supply from the *Earth* from below the pyramid shall be of the intensity of significantly increased value.
The summarised value shall be 1.6366 times more than the normal impact of gravitation.
And the concentration of the quantum impact of gravitation results in conflict.
The conflict is with increased *IQ* drive and intensity.

4C2 — The increased *IQ* demand of the pyramid is equal to: $IQ_{E+} = 1.6366\dfrac{c_E^2}{\varepsilon_E}$; | Which in the granite corresponds to: $IQ_{p+} = \dfrac{IQ_{E+}}{1.00852} = 1.46E+11$

4C3 — The aggregate quantum speed of the mineral will be not changing, as the space-time of the pyramid remains as it was, but the conflict is equivalent to the acceleration to speed *v*.
The equation gives an acceleration up to: $v = 236{,}360$ km/sec.

$$IQ_{p+} = \frac{c_p^2}{\varepsilon_p\sqrt{1-\dfrac{v^2}{c_p^2}}}$$

The conflict within the pyramid structure is equivalent to the speeding up of the mineral of the pyramid up to $v = 236{,}360$ km/sec. This conflict is a potential difference, which generates energy. The potential difference is independent on the height of the pyramid and for the "classical" $\varphi = 1.6189$ relation is measured as $U = 900\,mV$ value.

This potential is source of energy and generating electricity between the pyramid and the *Earth* surface.
In the case of a smaller pyramid, height of *150 mm*: it is $I = 0.28\,mA$; for *500 mm* height, it is $I = 0.5\,mA$. For the *146.5* m height of the *Great Pyramid of Giza* this would give much higher current value.

There was an important finding during the experiments: the internal temperature of the pyramid depends on the current, taken from the top of the pyramid.

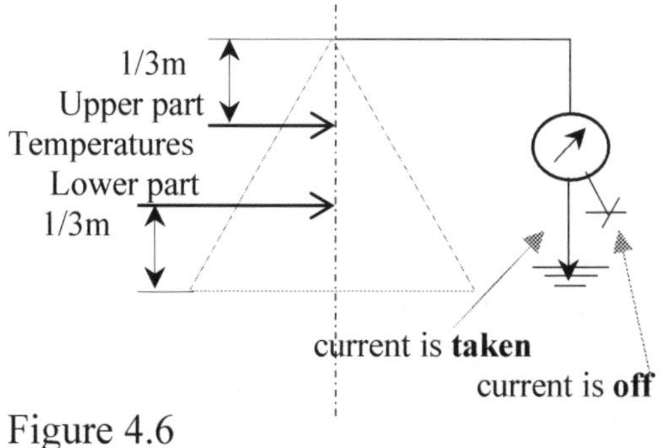

Figure 4.6

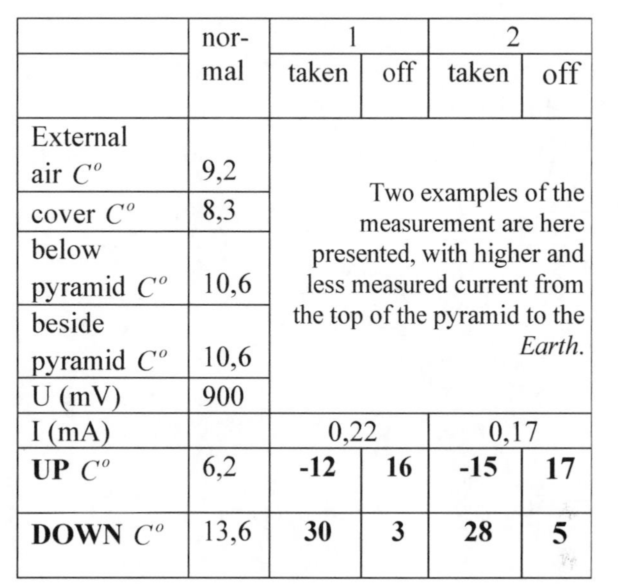

	nor-mal	1		2	
		taken	off	taken	off
External air C^o	9,2	Two examples of the measurement are here presented, with higher and less measured current from the top of the pyramid to the *Earth*.			
cover C^o	8,3				
below pyramid C^o	10,6				
beside pyramid C^o	10,6				
U (mV)	900				
I (mA)		0,22		0,17	
UP C^o	6,2	**-12**	**16**	**-15**	**17**
DOWN C^o	13,6	**30**	**3**	**28**	**5**

Fig. 4.6

"Current taken" means the moment when the top of the pyramid is connected to the *Earth* surface. "Off" means the moment when the connection is interrupted.

<u>The change of the internal temperature is immediate, without any time delay!</u>

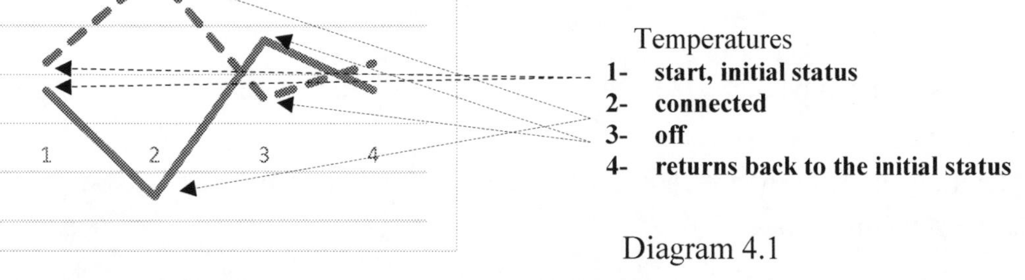

The internal temperatures are changing without any "classical" heat exchange!

The changes are immediate, without any time delay! The diagram is stretched just for better visualisation.

Diagram 4.1

Diag. 4.1

shows the temperature of the *upper* point;

is about the temperature at the *lower* segment.

The connection of the top of the pyramid with the *Earth* surface results in taking out electron processes from the upper part of the pyramid. Taking away electron processes decreases the conflict within the upper point of the measurement. Therefore the measured temperature becomes less.

The missing from the upper part electron processes will be supplemented from the lower segment of the pyramid; therefore the measured temperature at the point, one third from the bottom of the height of the pyramid is increasing.

The ruler is the continuity of the quantum impact of gravitation through the entire surface of the pyramid.

Once the connection to the *Earth* is off, the dynamism of the change is taking the temperature to the other end (line 3 on Diagram 4.1):

- The stop of the current from the upper part and the feeding from below results in conflict, which increases the temperature in the upper part;
- Once the supply from the lower part is not necessary anymore, the conflict is off in the lower section and the temperature comes down.

After the temperatures return back to the original status (points 4).

Further measurement examples are demonstrated on the following Diagram 4.2.

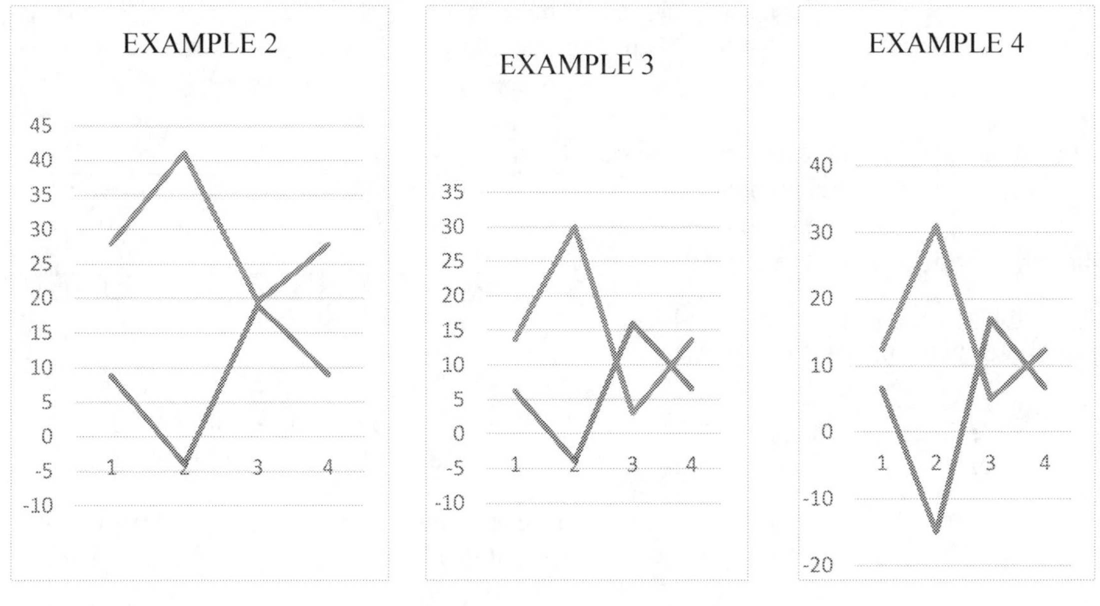

Diag. 4.2

Diagram 4.2 (a,b,c)

The changes in temperatures in the conditions and parameters of the experiment were about 10-15 ^{o}C.

The diagrams have been stretched for the better presentation of the case, but the changes themselves are immediate.

It is important to note, that the temperatures in the test pyramid, presented in Diagrams 4.1 and 4.2 were measured by two sets of thermocouples. The electromagnetic impacts of the operating thermocouples increase the dynamism of the change. Thermocouples strengthen the internal quantum conflict of the operation of the pyramid. This increased impact is detected even without taking away electricity from the pyramid.

S. 4.3.1

4.3.1 Pyramids are in quantum communication with each other

We need two or more pyramids for this experiment.

The distance between the pyramids has its impact on the intensity of the measured quantum signals, since signals propagate in quantum space, full with other signals.

The experiment is about taking away electricity from one of the pyramids and measuring the deceasing voltage, the internal potential in parallel of the other.

There is no any formal connection between the two pyramids.

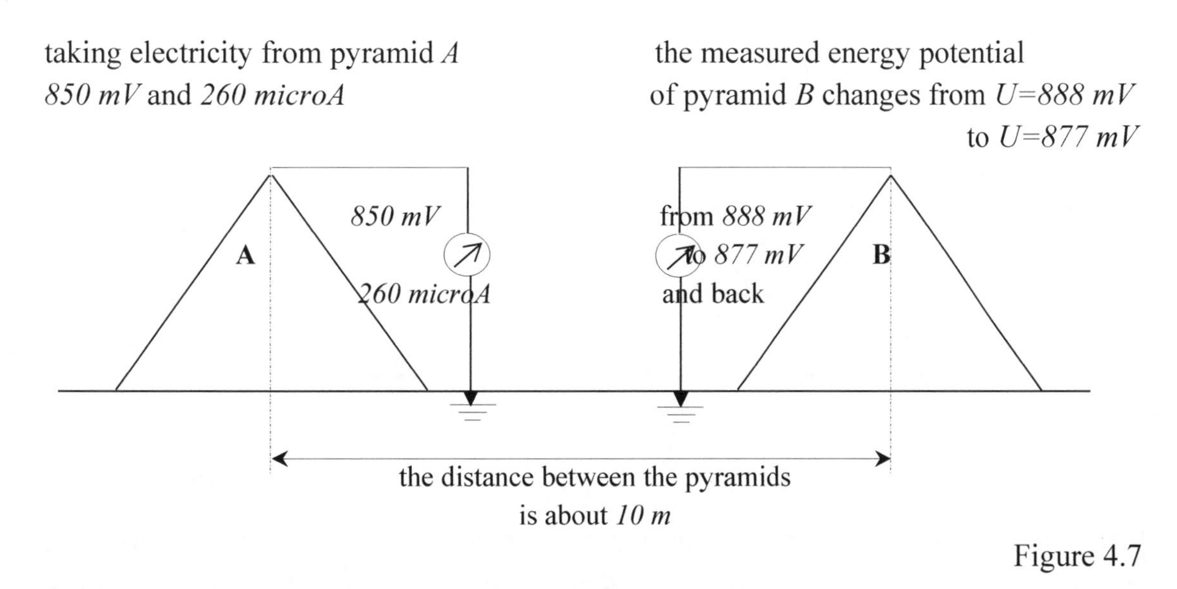

Figure 4.7

Fig. 4.7

The change in *B* is immediate and parallel with the change in *A*.
Once the current has been taken from *A*, the voltage in *B* changes from *U=888 mV* to *U=877 mV*. At the moment the draining from *A* stops, the potential of *B* returns to its original value.
The measured values of the voltage (and the current in *A*) are slightly decreasing in both pyramids in time, but the character of the change, the parallel impact remains the same.
The distance between the pyramids was about 10 m.
The change is not significant at all, since it is about just *10 mV*, but the most important finding in this case is the proof on the communication of the two pyramids in quantum space and time. This is only possible if the two pyramids have common space-time.

It can be stated in summary, that

- pyramids have their own internal quantum conflict, generated by the quantum impact of gravitation. This conflict can be influenced by external quantum impact, as the effect of the thermocouples proves it.
- pyramids communicate in quantum space, as the measured potential data prove it.

4.4
The quantum impact of *gravitation* on convex/concave surfaces

S. 4.4

The quantum impact of gravitation can also be demonstrated by other experiments as well:

A *convex* format plate, made from *Cuprum* process, positioned on the *Earth* surface as presented on Figure 4.8 on the next page, gives certain variety of temperatures all around the surface. The unique in the case is, that the temperature of all points of the convex surface are significantly higher than that of the *Earth* surface, which is $20.1\,^{o}C$ and the air temperature, which is $20.4\,^{o}C$.
The surface of the plate was protected from any external quantum impacts, like sunshine, heating or other kind.

The temperature at the top is not the highest. The highest measured values are on the cover with the highest curvature of the surface. (All data represent the average of the measured values on the surface at 6 spots of the same height and radius.)

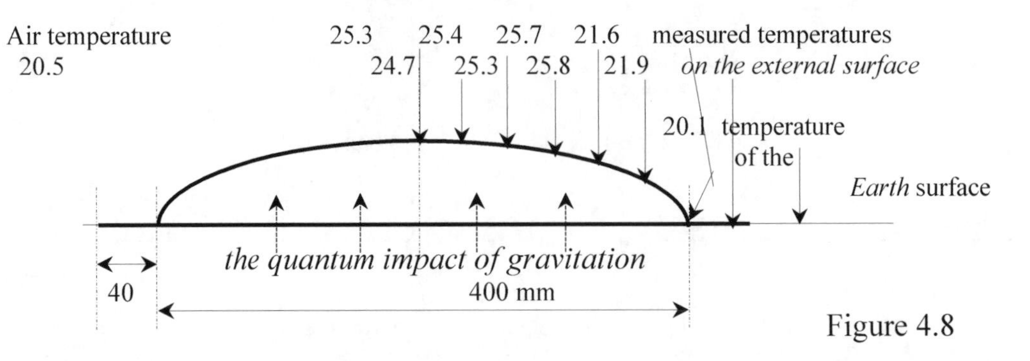

Fig. 4.8

Figure 4.8

The measured temperatures are consequence of the difference between the territory of the field, covered by the convex *Cuprum* process plate and the surface of the plate, similarly to the effect of the pyramid. The measured temperatures characterise the intensity of the conflict within the *Cuprum* process and the space below the plate.

Turning over the plate and making it *concave* from *convex*, the distribution of the temperatures becomes different.

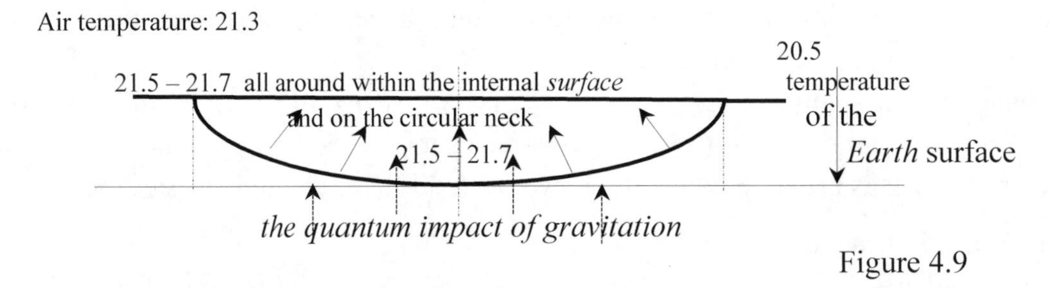

Fig. 4.9

Figure 4.9

The quantum impact of gravitation in this case results also in a certain distribution of the temperatures, higher than the temperature of the ground of the positioning, but the difference is of less value and the distribution is much more homogenous and closer to the temperature of the external air $21.3\,^{o}C$.

In the case of a *convex/concave* structure the conflict within the closed by the two halves is increased, but because of the increased but homogenous impact from below of the convex cover, the measured temperatures on the surface are more uniform.

Fig. 4.10

Figure 4.10

The measured temperatures, illustrated on Figure 4.11 on the next page are the average values measured in *six* segments of the two halves.

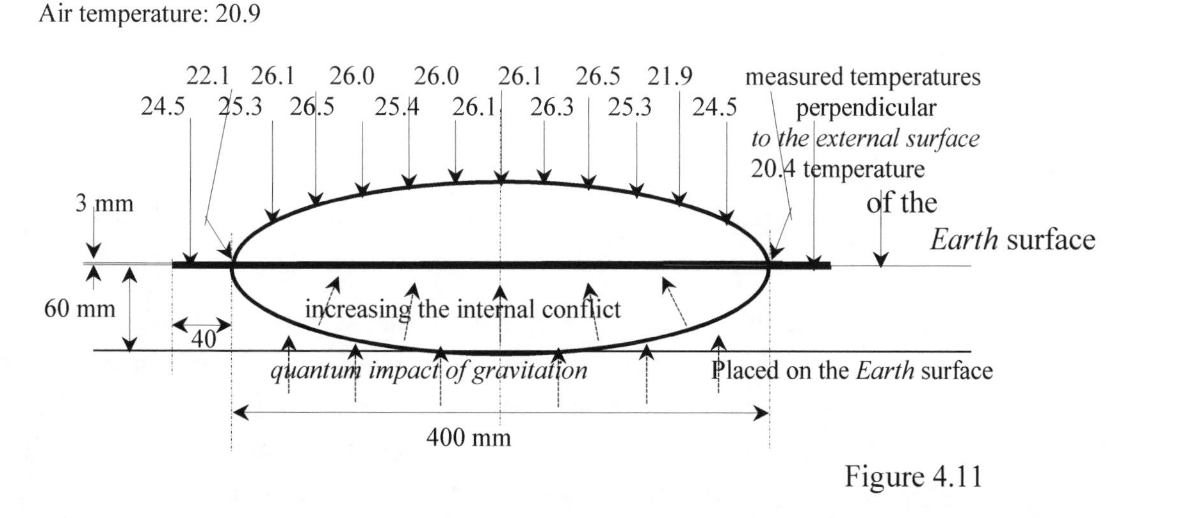

Figure 4.11

Fig. 4.11

The temperature of the *Cuprum* process in the experiment is higher than the temperature of the *Air* around the structure. It shall however be noted that all elementary processes with neutron process dominance (like the *Cuprum* process) have higher temperatures even in normal circumstances than the surrounding environment. The reason is the conflict at the anti-direction side, generated by the surplus of the anti-electron processes and the acting quantum impact of gravitation. The higher the periodic number is, the higher is the surplus and higher is the internal temperature of the elementary process. The *Uranium* elementary process for example has higher normal temperature than the *Cuprum* process.
The temperature of the flange is of higher temperature. It is part of the convex format from one side is open and on the other closed by the concave mantle.

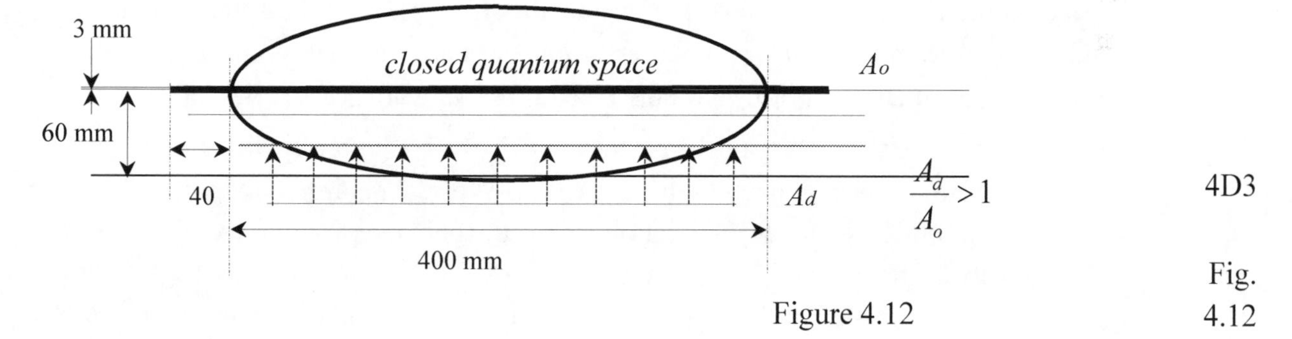

4D3

Figure 4.12

Fig. 4.12

The natural density of the quantum impact of *gravitation* one and the same and represent a constant value. If the configuration is with changing surface, not parallel with the *Earth* surface, the number of the quantum impacts from the surface of the subject to the quantum space is higher. The incoming simultaneous quantum impact of gravitation should obviously also be higher.

There is no difference in the quantum impact of gravitation. With reference to 4D3, the configuration of the surface is the one establishing the rate of the impacting quantum signals (and the temperature).

S. 4.5

4.5 Convex/concave structure with *Hydrogen* process inside

If the internal quantum space is filled with *Hydrogen* process, the *Hydrogen* process strengthens the conflict, as the *Hydrogen* process will be also in conflict.
If the concave/convex structure in addition is rotating, the impact of the conflict is even more efficient.

Rotation means constant acceleration at each point of the radius of the construction. Rotation increases the temperature within the elementary process of the structure and inside the convex/concave structure. The temperature is increasing towards the periphery and the deformation of the curved plate compensates the impact without the damage of the structure. The increase of the temperature is consequence of the conflict of the increasing intensity of the electron processes.

But the electron processes are results. Results of the proton processes. If the intensity of the elementary process becomes increased, than the intensities of the proton and the neutron processes in parallel are also increased. The acting electron process is the continuity of the proton process at the standard to the elementary process intensity just the acceleration is increasing its value – as the acceleration is increasing the intensity of the proton process as well. (The neutron process is driven by the electron process. If the electron process is of increased intensity, the neutron process is of increased intensity as well.)

The proton processes as events are similar for all elementary processes; just the quantum speed values and the duration of the events are different. And the proton processes are ending at certain intensity values. The time systems at that point for all electron processes are equal: $\lim t_i = \propto$, with the quantum speed values are different.

So, the quantum drive values of the electron processes are certainly different!
And the electron process quantum drive is the drive of the neutron collapse.

And the conflict on the anti-direction side is the one determining the intensity of the drive of the electron process. As all proton processes shall follow the intensity of the anti-proton collapse: $\frac{c_x^2}{\varepsilon_{x-}} = const$; (4E1)

The result of the quadrat of the quantum speed and the intensity coefficient of the electron process for all elementary processes is constant: $c_x^2 \cdot \varepsilon_x = const$. (4E2)

The intensity of the proton process will be kept within standards by the generation and the surplus of the anti-electron processes.

The increased intensity of the direct side results in increased collapse and in extra surplus of the anti-electron processes for keeping the proton process at standard level. The quantum speed belongs to the elementary (proton) process.

$$\frac{dmc_x^2}{dt_i\varepsilon_x\sqrt{1-\frac{v^2}{c_x^2}}}\left(1-\sqrt{1-\frac{(c_x-i_x)^2}{c_x^2}}\right); \quad \text{(4E3)}$$

The higher is the radius, the higher is v, the speed of the acceleration of the rotation.

The quantum speed of the elementary process does not change as the elementary processes of the construction. It remains the same, unchanged. The generating in the elementary process of the construction anti-electron process surplus will be in conflict with the quantum impact of gravitation.

And the generating surplus is: and for the structure: $\varepsilon_{x-} > 1$

$$\Delta e_{-} = \frac{dmc_x^2}{dt_i \varepsilon_x \sqrt{1-\frac{v^2}{c_x^2}}} - \frac{dmc_x^2}{dt_i \varepsilon_x} = \varepsilon_{x-} \cdot e_{-}\left(\frac{1}{\sqrt{1-\frac{v^2}{c_x^2}}} - 1\right); \quad \text{as } \varepsilon_{x-} = \frac{1}{\varepsilon_x} \qquad \text{4E4}$$

If the convex/concave structure has *Hydrogen* process inside and this gas content is accelerated additionally within the construction, the surplus is generating at the direct side of the gaseous elementary process.
Surpluses at the anti-electron process side of the structure plus the electron process side of the *Hydrogen* content inside generate in parallel quantum impacts.

The anti-electron process control keeps the proton process in value, but the electron process as above, has its intensity increase. The intensity increase is generating at the anti-electron process side, which quantum impact will be conflicting with the quantum impact of gravitation. Plus to all this, the accelerating inside *Hydrogen* process will also be in conflict with the increased anti-electron process quantum impact of the elementary process of the construction and with the quantum impact of gravitation.
The conflict with the quantum impact of *gravitation* is generating lifting effect!

The lifting effect of the conflict depends on the speed value of the acceleration.
Motion generates quantum impact; quantum impact generates motion. As practical proof of relativity!

4.6 *Earth's* expansion

S 4.6

The *Hydrogen* process has its accumulating character. The classical *turn-around-global-inflexion-plasma* process has its immediate change character. All other intermediate statuses are between the two.

$$\frac{c_x^2}{\varepsilon_{x-}} = const \qquad \text{4F1}$$

The anti-processes follow the rule and the presented equal quantum drives of all anti-electron processes and result in similar proton processes, happening in different space-times.

The speed of the quantum communication and the intensity of the electron process, the quantum drive are the keys of the cooling process. The intensity of the anti-neutron process of the plasma inflexion is of infinite high value. The expansion of the anti-neutron process results in infinite large number of anti-electron processes.

$$\frac{dmc_x^2}{dt_n \varepsilon_x}\left(1 - \varepsilon_x \sqrt{1-\frac{i_x^2}{c_x^2}}\right) = \frac{dmc_x^2}{dt_n \varepsilon_x} - \frac{dmc_x^2}{dt_i}; \qquad \text{4F2}$$

The intensity of the anti-neutron process expansion shall correspond to the intensity of the neutron collapse, which is driven.

In the case of the *plasma* process, the drive of the neutron process is of infinite high intensity. The electron/neutron process matrix of the *Hydrogen* process is collapsing for infinite short time. The characteristics of the quantum drive are the intensity of the electron process and the quantum speed.

4F3 $$dt_i = \frac{dt_n}{\sqrt{1-\frac{i_x^2}{c_x^2}}}$$ The *end point* of the anti-neutron expansion, with reference to the above is:

4F4 $$\frac{dmc_x^2}{dt_i}$$ The intensity coefficient of the anti-drive, shall correspond to ε_{x-} as $c_x^2 \cdot \varepsilon_x = const$.

4F5 $$\varepsilon_{x-} = \frac{1}{\varepsilon_x}; \quad \varepsilon_{x-} = \frac{\varepsilon_{n-}}{\varepsilon_{p-}};$$

The quantum drive of the anti-proton process is: $$\frac{dmc_x^2}{dt_i\varepsilon_{x-}}\left(1-\sqrt{1-\frac{(c_x-i_x)^2}{c_x^2}}\right)$$

The anti-neutron expansion shall be ending with dt_i time count. With reference to 4F2, this corresponds to $i_x = \lim a_x \cdot \Delta t = c_x$ status, the start of the anti-electron process. But the number and the intensity of the anti-electron process as drive shall meet the elementary balance, specified in Section 1.1. With reference to 1C8,

Ref. S.1.1 1C8

4F6 the quantum membrane of the anti-electron process establishes the intensity of the electron process: $$\frac{dmc_x^2}{dt_i\varepsilon_x} = n_x \frac{dmc_x^2}{dt_i\varepsilon_{x-}};$$

Therefore the anti-neutron process of the plasma generates infinite large number of anti-electron processes:

4F7 $$n_x = \frac{\varepsilon_{x-}}{\varepsilon_x}; \quad \text{as } \lim \varepsilon_x = 0 \text{ and } \lim \varepsilon_{x-} = \propto = \lim n_x$$

The *global turn-around inflexion of the plasma* generates infinite large number of anti-electron processes. The anti-electron processes are in conflict. The subsequent high number of the sub-plasma stages results in additional anti-electron processes and the conflict is escalating. The plasma state is about the intensity conflict of the anti-electron processes.

With the progress of the elementary evolution, the intensity of the anti-neutron expansion is decreasing step by step. As the intensities of the quantum drives of all anti-proton processes are of equal values, the intensity coefficient of the anti-electron process is decreasing!

There are here *two* consequences to be noted:

- the conflict is acting while the elementary evolution is going on;
- the conflict is impacting the developing external quantum membrane, resulting in *gravitation.*

  The intensity of the quantum impact of gravitation is decreasing with the progress of elementary evolution, as the generation of the anti-electron process surplus is getting less and less.

The decreasing tendency of the generation of the anti-electron process surplus means, the *proportion* of the anti-proton/proton inflexions – during the progress of the elementary evolution – is increasing.

With the decrease of the anti-electron process surplus, the "quantum pressure" of the quantum membrane at the anti-side is decreasing and the intensity of the electron process is getting less and less with the progress of the elementary evolution; the duration of the elementary processes is extending.

$$IQ_x = \frac{c_x^2}{\varepsilon_x}; \text{ and } \frac{dmc_x^2}{dt_i \varepsilon_x}\left(1-\sqrt{1-\frac{(c_x - i_x)}{c_x^2}}\right) = n_x \cdot q \qquad \text{4G1}$$

With the progress of the elementary evolution the number of the impacted quantum becomes less and less.

The quantum drive of the plasma is of infinite high value.

The increasing time count means reduced intensity, higher value of the intensity coefficient, and decreasing quantum speed in 4G1. The "cooling" of the plasma goes on and the generating elementary processes of the evolution shall correspond to the intensity of the plasma. The subsequent elementary statuses therefore are generating in increasing numbers with the increasing number of electron processes. Ref. 4G1

The quantum drive in 4G1 is the characteristic of the elementary process.

The acting time system of the electron process is the function of the intensity coefficient:

$dt_x = dt_i \cdot \varepsilon_x$ - the acting time system with the evolution is increasing, as the value of ε_x is increasing; 4G2

$dt_{plasma} < dt_x < ... < dt_H$ the acting time system is increasing during the elementary evolution. The $c_x^2 \cdot \varepsilon_x = const$ means the permanent decrease of c_x the value of the quantum speed in the elementary processes. 4G3

The intensity gap between the last, operating in full cycle *Helium* process and the *Hydrogen* process is category of infinite high value. (The intensity coefficient of the *Hydrogen* process is $\lim \varepsilon_H = \propto$, the intensity coefficient of the *Helium* process is $\varepsilon_{He} = 1.014$.) But there should be no gap here between the *Helium* and the *Hydrogen* processes. The elementary evolution continues, as the vegetation on the *Earth* surface clearly proofs it to us, even with not fully completed elementary processes:

> The *Earth* core is full of elementary processes with not completed elementary cycles. The completion needs quantum communication. The communication is initiated by the processes themselves. And the request has been met by those 8 elementary processes with electron process surplus and proton process dominance.

With reference to the quantum drive formula in 4G1, the decrease of the quantum speed and the decrease of the intensity (increase of the coefficient) of the electron process mean: Ref. 4G1

- the original quantum impact of the plasma with n_x impacted quantum can only be ensured, if the number of the participating in the quantum communication electron process *blue shift* impacts (x) is increased.

Otherwise the two sides of the equation here below cannot be equal.

$$x\frac{dmc_x^2}{dt_i \varepsilon_x}\left(1-\sqrt{1-\frac{(c_x - i_x)}{c_x^2}}\right) = n_x \cdot q; \qquad x\frac{c_x^2}{\varepsilon_x} \equiv n_x \cdot q \qquad \text{4G4}$$

The intensity of the electron processes (the *IQ* value) is decreasing, their number however is increasing.

The cooling of the plasma is only possible if keeping the balance with the quantum impact of the plasma of n_x energy quantum. This means, with the cooling of the plasma – the increase of the number of the elementary processes, the expansion is natural consequence.

The *plasma* process has infinite number of elementary cycles, but all cycles lose infinite large anti-electron processes surplus on the quantum impact of gravitation. The sub and sub-sub plasma statuses and the elementary processes of the core have less and less cycles, with less and less anti-electron process "losses" on gravitation.

In line with the decrease of the absolute value of the intensity of the expansion (as the time count is increasing and quantum speed values are decreasing), the absolute value of the intensity of the neutron collapse is also decreasing.

The infinite high density value of the plasma process status is reducing to infinite less values of intensity and increased time counts of the elementary processes. While the number of the parallel cycles becomes radically decreased, the reduction of the operating in parallel elementary processes is moderate, as the loss on gravitation is moderate.

Ref. QISM 1N5

4G5 With reference to the Section 1.7 of *The Quantum Impulse and the Space-Time Matrix* and formula 1N5 $dl_x = A \cdot f\left(\frac{1}{c_x}\right)$;

where A – is a constant, relating to all elementary processes in equal way.

4G5 above is coming from the rearrangements of the $dl_x = c_x(dt_i \cdot \varepsilon_x)$ formulation, the characteristic of the density of the event: *The less the quantum speed value is, the larger is the distance the quantum communication needs*.

This formula in 4G5 is not about space-time.

4G5 in fact is the opposite to the space-time dimensions. Space-times of increased quantum speed values are larger and larger while the quantum communication needs less and less time count. With the decrease of the quantum speed value the time count of the communication here is increasing:

Ref. QISM 5C5 5C7

4G6 With reference to the Section 5.2 of *The Quantum Impulse and the Space-Time Matrix* and formula 5C5, 5C7 $f(c_x) = f(n_x) = f\left(\frac{1}{\Delta t_x}\right)$;

The lengthening of the relative distance dl_x in 4G6 and the increasing of the relative time dt_x mean = the *Earth* is expanding!

If the intensity on the direct side is decreasing, the intensity of the anti-direction is going up; meaning, the intensity coefficient of the anti-electron process ε_{x-} is of less and less value. The decreasing impact of ε_{x-} on the generation of the anti-electron process surplus:

4G7 $n_s = n_x(\varepsilon_{x-} - 1)$,

and the decreasing value of the quantum speed in parallel means the decreasing tendency of the global quantum impact of *gravitation*.

In summary:

With reference to 4G5 and 4G6, with the cooling of the plasma the relative distance demand of the elementary evolution is increasing. This means expansion. Ref. 4G5 4G6

The intensity and the quantum speed of the electron processes are decreasing, but the variety (the number) of the elementary processes is increasing. It results in permanent *mechanical impact of gravitation* with decreasing intensity. At the same time, the quantum impact of the developing anti-electron process surplus is decreasing and this way the intensity of the *quantum impact of gravitation* is decreasing.

4.7 Specific conditions of *Earth's* expansion

S. 4.7

Elementary processes with electron process *blue shift* surplus and conflict on the *Earth* surface establish liquid (water, hydrocarbons), gaseous (*H, He, Ni, O*) elementary statuses, and create specific intensity conditions in elementary processes (*C, Si, S, Ca*).

It means the continuity of the quantum impact within the core and above the core.
The elementary evolution results in global mechanical impact, the expansion of the *Earth* surface. There might be differences in the local impacts. It is function of the elementary composition of the core.

With the decrease of the quantum speed and with the increase of the intensity coefficient, the duration of the quantum impact of the elementary process becomes of increased value:	$\frac{c_x^2}{\varepsilon_x} = IQ_x$: less IQ value shall be acting for longer: $\Delta t_x \approx IQ$	4H1 4H2

The cooling process of the plasma results in the reduction of the quantum speed and the increase of the time count. Higher time count with less value of the quantum speed means longer distance and increased time count of the quantum communication.
Gravitation means expansion!

$$\frac{dl_{x1}}{dt_i} \neq \frac{dl_{x2}}{dt_i}; \text{ and } \frac{dl_{x1}}{dt_i} > \frac{dl_{x2}}{dt_i} \text{ it might be realised or might be not.} \quad 4H3$$

$$IQ_{x2} = n_{x2} IQ_{x1}, \text{ which means: } \frac{c_{x2}^2}{\varepsilon_{x2}} = n_{x2}\frac{c_{x1}^2}{\varepsilon_{x1}} = \frac{c_{x1}^2}{\varepsilon_{x1}\sqrt{1-\frac{v^2}{c_{x1}^2}}}; \text{ and } IQ_{x2} > IQ_{x1} \quad 4H4$$

In the case of differences in the elementary composition of the core, the difference in the values of the quantum speed influences the intensity of the expansion. Higher quantum speed means the distance of the quantum communication is of less value and shorter time count. As consequence, the intensity of the mechanical impact, the expansion itself of the core with elementary process of less quantum speed is higher!!

The *expansion-gravitation* also means events and impacts on the subject above the *Earth* surface. In the case of higher expanding intensity of the gravitation impact, the acceleration is higher.

As the mechanical impact of gravitation is also function of the positioning, it can happen that in the case of significant difference in the mineral composition of hill, region, the summarised impact of the expansion is different than the everyday practice:
Subjects may move free into uphill direction:

4H5

- with reference to 4G3: $\frac{dl_{x1}}{dt_i} > \frac{dl_{x2}}{dt_i}$;
- the intensity impact of the composition takes over the impact of the positioning;
- there is no difference in the time count, as both impacts belong to dt_i time system.

It is important to note, that all these results and findings on the subjects of *Earth* expansion and of elementary evolution from the *plasma* to the *Hydrogen* process are in harmony with everyday practice and with conventional experience.

S. 4.8

4.8
Ice in its many formats

The difference between the elementary processes with electron process surplus is the value of the quantum speed. The ones with less quantum speed than the quantum speed of the *Earth* surface become gaseous, as gravitation speeds them up and create additional conflict, which turns them into gas aggregate status. The ones with higher quantum speed cannot be impacted as their quantum drive is of higher value. So, the aggregate status is not just the function of the status of the electron process.

Water on the *Earth* surface represents the communication of the *Oxygen* and the *Hydrogen* processes at the temperature of the status of the elementary evolution in the *Earth* core.

The elementary quantum speed of the *Oxygen* process is a certain one, less in value than the speed of light on the *Earth* surface. The quantum speed of the *Hydrogen* process is infinite low. The *Earth* core has its influence on the electron process *blue shift* conflict of the water, as the elementary evolution dictates a certain conflicting stage – temperature – within the core.
So, the elementary communication between the *Oxygen* and the *Hydrogen* processes becomes increased. The two together plus the intensity impact of the global elementary evolution establish an increased intensity, with quantum speed value of the water less than the speed of light.

> The *Hydrogen* process is of infinite low quantum speed and infinite low intensity, so in natural circumstances the temperature of the *Hydrogen* process is absolute zero, meaning = no conflict!
> The *Oxygen* process has its certain quantum speed value, less, but close to the speed of light. The *Oxygen* process has its conflicting status.
> The quantum impact of gravitation increases the value of the quantum speed of both elementary processes to the speed of light above the *Earth* surface. And this way they become gaseous.

The cooling impact to the water decreases the increased intensity effect of the quantum impact of gravitation and of the elementary processes, received in the *Earth* core. Cooling means taking conflict away from the water, meaning: the electron process *blue shift* surplus of the water is externally used. Consequence = decrease of the temperature. Water loses on the intensity of its *blue shift* conflict. Taking away electron process *blue shift* impacts, the *Oxygen* and the *Hydrogen* processes of the composition remain the same, but the internal conflicting effect is weakening.

At around $4\,^{o}C$, water reaches its specific, liquid aggregate status. All intensity surpluses received from gravitation and much more than the intensity impact of the elementary evolution has been taken away. (As within the core the temperature is higher than $4\,^{o}C$.) The water at this temperature still has its electron process surplus, since the *Oxygen* process is still with conflicting electron process surplus and the *Hydrogen* process is always with surplus. The aggregate quantum speed value is still less than the speed of light on the *Earth* surface. The water is still of liquid status.

With the further cooling, the *Hydrogen* process is the one, which is losing on its quantum speed value, result of the impact of the elementary evolution in the core. (The quantum speed of the *Hydrogen* process at $\lim t_H = 0$, corresponds to $\lim c_H = 0$.) The resulting less aggregate quantum speed of the water means space-time of less intensity. Space-time of less intensity corresponds to increased time count, less efficiency, in conventional terms to increased volume. At $0\,^{o}C$ the aggregate quantum speed value result in solid status.

Ch.5

5
The Standard Model with process based argumentation

The experimental findings of the standard official elementary model are well in line with the process based approach. This harmony is not just about

5A1 the sphere symmetrical expanding acceleration of the <u>proton process</u> with increasing speed value of: $e_p = \frac{dmc_x^2}{dt_p}\left(1-\sqrt{1-\frac{v_x^2}{c_x^2}}\right)$;

5A2 the sphere symmetrical expanding acceleration of the <u>electron process</u> at constant $\lim i_x = c_x$ speed value: $e_e = \frac{dmc_x^2}{dt_i \varepsilon_e}\left(1-\sqrt{1-\frac{(c_x-i_x)^2}{c_x^2}}\right)$;

5A3 and the sphere symmetrical collapse of the <u>neutron process</u>, driven by the electron process: $e_n = \frac{dmc_x^2}{dt_n}\sqrt{1-\frac{(c_x-i_x)^2}{c_x^2}}\left(\sqrt{1-\frac{v_x^2}{c_x^2}}-1\right)$;

but also about all measured impulses and particles of the standard model.

For proving this, we are going to assess all measured particles of the ***strong*** and the ***weak interrelations*** in line with the process based approach in the following sections.

The *proton* and the *neutron* processes are composed from <u>three-three quark processes</u>:

Proton process	Neutron process
1. the *top-charm-up* process is in balance with	1. the *down-strange-bottom* process is in balance with
2. the *down-strange-bottom* process;	2. the *top-charm-down* process;
3. plus the *top-charm-up* process is without balance	3. plus the *down-strange-bottom* is without balance

Ref. S. 1.6

With reference to Section 1.6, the quark processes within the proton and the neutron and within the anti-neutron and the anti-proton processes represent the balance of the elementary communication.

The quantum communication happens within the processes of hadrons of the *strong interrelation*; meaning: (1) within the same hadrons, and (2) between the hadrons as well.

The basics of the communication are the adequacy of the space-times, established by the quantum speed of the communication.

The electron and the anti-electron processes, as processes of the *weak interrelation* are not just the quantum drives of the communication, but also the ones establishing the space-time – by the quantum speed of the drive – of the driven process. Therefore there is no surprise at all having quark processes of different directions (meaning: expansion versus collapse) within the proton, the neutron and the anti-neutron and anti-proton processes.

Diagram 5.1 below shows the elementary process with the acting components.

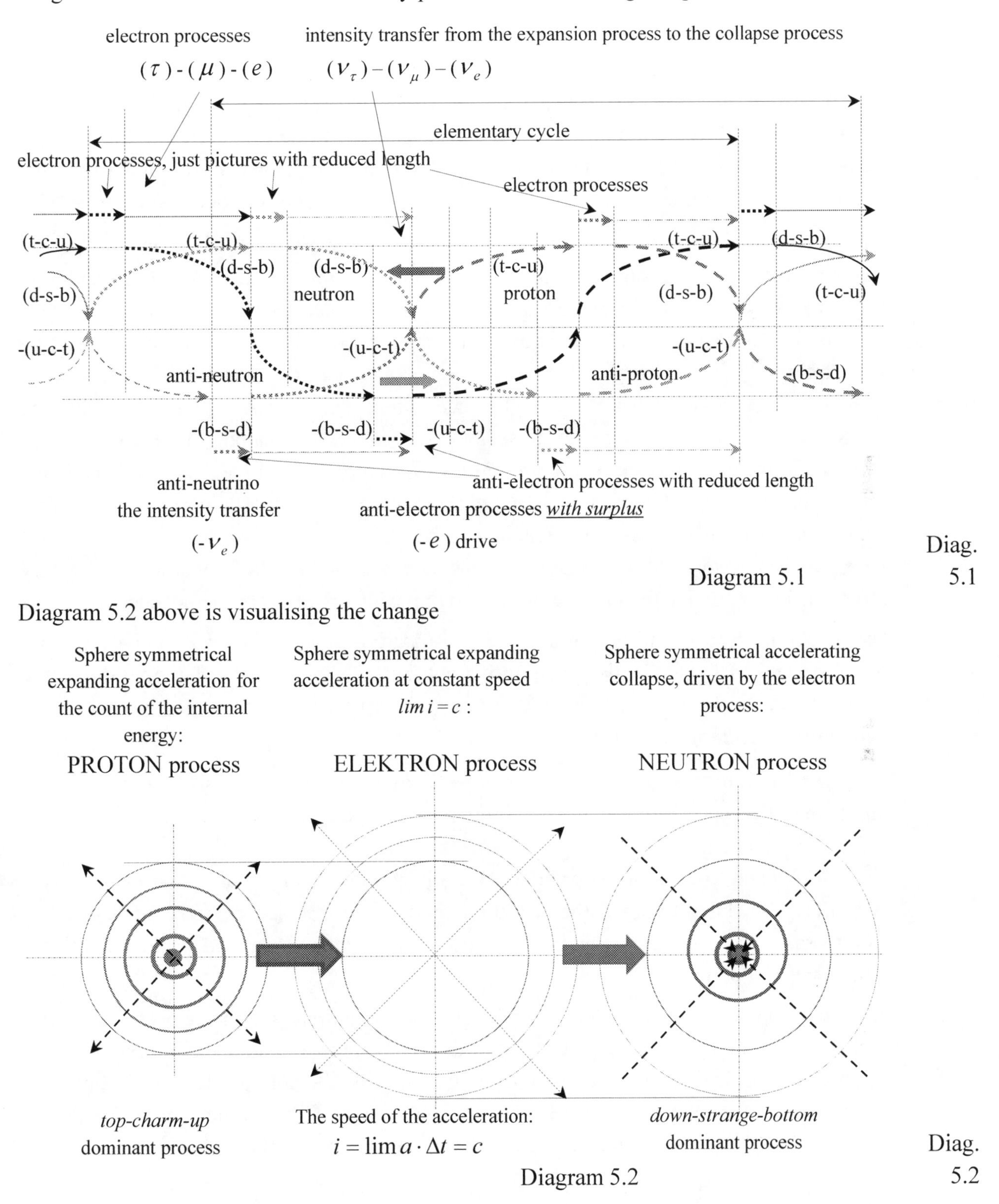

Diagram 5.1

Diagram 5.2 above is visualising the change

Diagram 5.2

The proton process is the ***decrease*** of the energy intensity, a sphere symmetrical expanding acceleration, represented by the *top-charm-up* quark process chain, from 170 GeV/c^2, the *top*, to 2.3 MeV/c^2, the *up* quark via 1.3 GeV/c^2, the *charm*.

The neutron process is the *down-strange-bottom* chain, a sphere symmetrical accelerating collapse, the ***increase*** of the intensity from 4.8 MeV/c^2, the *down* up to 4.8 GeV/c^2, the *bottom* via 95 MeV/c^2, the *strange*.
The measured mass/energy values, the impacted number of quantum by the *quark processes* are well representing the changing intensity of these events in the directions of the decrease of the intensity of the process (proton process - the expansion) and also in the direction of the increase of the intensity of the process (the neutron process - the collapse).

The proton process as sphere symmetrical expanding acceleration is also contains a chain of *down-strange-bottom* quark processes with sphere symmetrical accelerating collapse. The same way, the neutron process as sphere symmetrical accelerating collapse is also contains a chain of *top-charm-up* quark processes with sphere symmetrical expanding acceleration.

Ref. S. 1.6

These quark processes are however in balance within the *hadrons*, in the proton and the neutron processes; they are parts of the same space-time, as explained in Section 1.6. The third quark process is the reason the proton process is representing the *acceleration* and the neutron process the *collapse*: an *expansion* line within the proton process and a *collapse* line within the neutron process are without compensation.
This is the reason/tool/instrument of the internal and/or the external elementary communication: finding the balance of the two opposite directions of quark processes.

This is similar for the anti-proton and the anti-neutron processes as well, as Diagram 5.1 demonstrates it. In the anti-direction the anti-neutron process is the one in sphere symmetrical expanding acceleration and the anti-proton process is the one in sphere symmetrical accelerating collapse.

The *top, charm, up, down, strange* and *bottom* flavours of the *quarks* as we conventionally call them are the specific intensity stages of the expansion and the collapse in the direct and in the anti-directions.
The *u-u-d* characteristics of the proton process is coming from the fact, that these are the intensities, which make not just the measurement possible, but also stay for long in effect.
It is the same way for the neutron process, where the composition is *d-d-u* and the collapse is the dominant part of the process.

In the anti-processes of the elementary cycles the directions are the opposite: the anti-neutron process is repeating the intensity of the neutron process after the inflexion, just in the direction of the expansion; the anti-proton process is the collapse leading to inflexion before the start of the proton expansion. The intensity of the expansion of the proton process is repeating the accumulating intensity of the collapse of the anti-proton process.

Ref. S.2

The *inflexions* between the neutron/anti-neutron and the anti-proton/proton processes connect the two directions. There are 3 parallel cycles within the elementary processes.
The electron and the anti-electron processes within the two of the three cycles drive the corresponding *(d-s-b)* and *(u-c-t)* collapses. The remaining electron and anti-electron process surplus, if any, are for the internal and the external elementary communication.

The neutron process is neutral indeed, as ε_x the intensity coefficient of the electron process establishes the intensity of the collapse, the communication between the proton and the neutron processes:

$$\varepsilon_x = \frac{\varepsilon_p}{\varepsilon_n}\sqrt{1-\frac{(c_x - i_x)^2}{c_x^2}} = \frac{\Delta t_n}{\Delta t_p};$$ 5A4

The communication of the strong and the weak interrelations (interactions) establishes the form of the elementary process: the expansion (the proton process) is establishing the quantum speed; the electron process represents this quantum speed, (as part of the extension) and extends the impact of the space-time of this quantum speed (as the drive of the collapse). The anti-process brings the process back to the beginning as discussed in Section 2 and the next cycle starts again with the direct process.

Ref. S.2

5.1
The space-time of the proton and the neutron processes

S. 5.1

The proton and the neutron processes have their own specific relation.
The anti-proton/proton inflexion establishes the quantum speed of the communication of the elementary process. The electron process drives the collapse in line with the quantum speed of the expansion. In other words: the intensity of the expansion establishes the electron process, which drives the collapse. With reference to 1K5 and 1K6, the balance is generating between the expansion and the collapse.

The frequency of the expansion is in its natural step by step decrease during the expansion. The reason of the decrease is to balance the space-time of the elementary process, as there is another process within the same space-time, which however is about the increase of the intensity – the collapse.

$$\frac{dmc_x^2}{dt_p}\left(1-\sqrt{1-\frac{v^2}{c_x^2}}\right);$$

$$\frac{dmc_x^2}{dt_i\sqrt{1-\frac{v^2}{c_x^2}}}\sqrt{1-\frac{(c_x-i_x)^2}{c_x^2}}\left(\sqrt{1-\frac{v^2}{c_x^2}}-1\right);$$

Ref. 1K5 1K6

The *t-c-u* and the *d-s-b* quark lines in the proton and neutron processes, and also the *(b-s-d)* and *(u-c-t)* quark lines within the anti-neutron and anti-proton processes demonstrate they are parts of one and the same space-time. There is no expansion without collapse. The intensity of the collapse initiates the transfer of the intensity cover.
The proton process has its "internal" *d-s-b* collapse, driven by the electron process, and has its *t-c-u* expansion without response. This is the reason the proton process is expansion-dominant. The neutron process – being part of the same space-time – has its collapse response to the expansion within the proton process and has its balanced collapse-expansion pair as well. The key is the speed value of the quantum communication. This is the one establishing the space-time of the processes, independently on the specifics of the processes. The space-time of the collapse is established by the quantum speed value of the drive. This is the logic of the communication, whatever is the "separation" of the processes in space.

In the case of number of acting elementary processes, the intensity, the *IQ* of the drive is the one establishing the priority of the communication. The highest *IQ* value has its priority. Once however the priorities of the space-times have been established, all quantum drives are acting accordingly.
This is the reason the proton and the neutron processes have their *t-c-u* and *d-s-b* quark lines and the anti-neutron and anti-proton processes their *(d-s-b)* and *(t-c-u)* quark lines.

The quantum communication between the elementary processes is very much influenced by the existing conflicts (the distance within our space-time) between the communicating elementary parts. Therefore the direct communication has the highest efficiency of the elementary quantum communication.

S. 5.2

5.2. *Intensities* instead of mass

Our conventional view identifies the "mass" of the protons and the neutrons. These "mass" values however are rather the intensities of the quantum impacts of the ongoing processes, as their measured values also well demonstrate it. The measured impacts of the processes give the explanation on the discrepancy of the "masses" of the composing quark processes of the hadrons. The discrepancy is obvious: the summarised "mass values" of the composing quarks are much higher than the "mass values" of the proton and the neutron processes themselves, which are around $1.67 \cdot 10^{-27}$ kg, or 939 MeV/c^2 respectively.

The "measured mass" of the hadrons is no other than the intensity impact of the change. The three quark processes have different intensities (and durations) in their expansion and collapse. The hadrons, taken conventionally as particles represent the integrated intensity impact of the changes of the quarks (the "mass"), representing in fact the intensities of the longest, dominant processes.

The sphere symmetrical ***expanding acceleration*** is driven by the internal energy intensity of the process. The sphere symmetrical ***accelerating collapse*** needs drive.
All *down-strange-bottom* quark processes on the direct elementary side and all *up-charm-top* processes on the anti-side need driving impacts. These drives are the electron and the anti-electron processes. But the electron process as specific expanding acceleration has also three "flavours": *tau, muon* and *electron* with drive impact intensity values of

the τ (tau) stage with 1.7 GeV/c^2 - the drive for the shortest period,

the μ (muon) stage with 100 MeV/c^2 impact and

the *e* (electron) stage with 0.5 MeV/c^2 the longest impact.

The question is, how can a proton process *up* quark in its expanded 2.3 MeV/c^2 state produce an electron process with the intensity of 1.7 GeV/c^2?
The same way could the question be posed: how can the intensities, as above, drive the collapse of the *down* quark process with 4.8 GeV/c^2 intensity?

The answer is coming directly from the process concept:
The measured intensities are the locally measured values of a process which works out a certain intensity capacity. In the case of the electron process the intensity of the drive is generated by the quantum membrane of the anti-electron process. In the case of the proton-neutron communication, there is a constant intensity transfer between the expansion and the collapse, which is increasing the intensity of the driven process.

The intensity (the measured mass) of the electron process is function of the internal *quantum membrane* of the elementary process, established by the anti-electron process surplus of the elementary process.
The effects of the electron and the anti-electron processes belong to each other. The surplus at one side is impacting the intensity on the other.

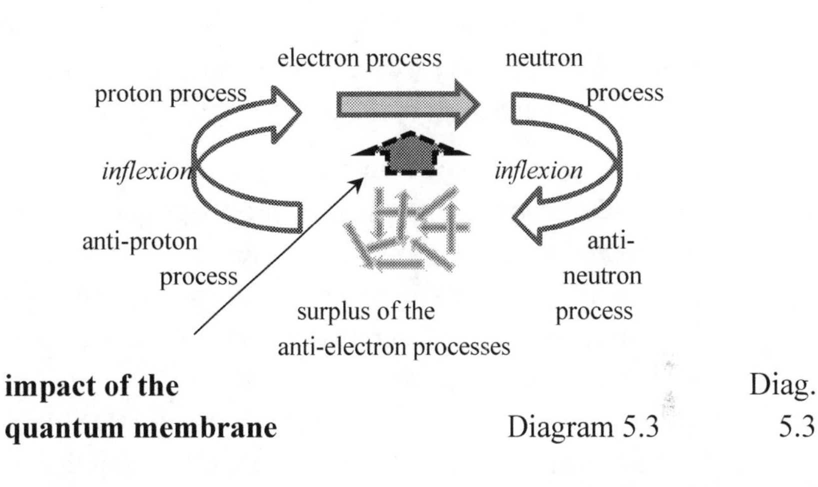

impact of the quantum membrane Diagram 5.3

Diag. 5.3

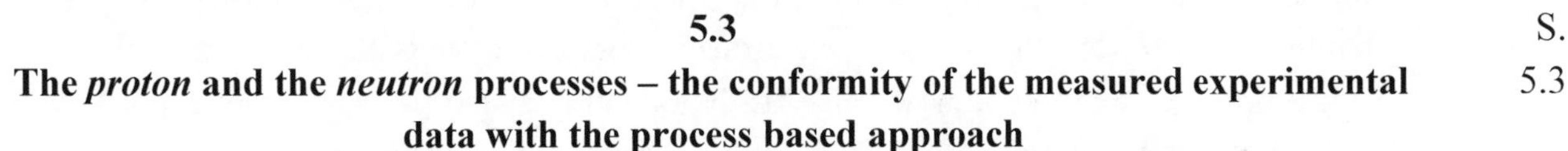

5.3 The *proton* and the *neutron* processes – the conformity of the measured experimental data with the process based approach

S. 5.3

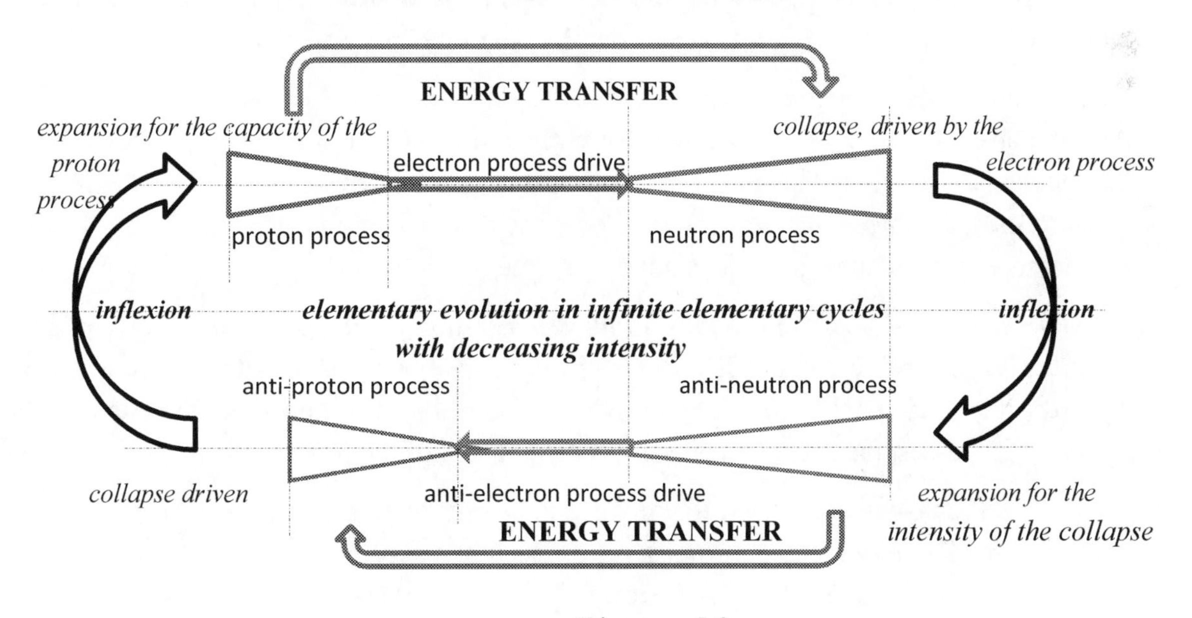

Diagram 5.3

Diag. 5.3

The *(t-c-u)* and the *(b-s-d)* processes and the electron and the anti-electron processes, including the *muon,* the *tau* and the *anti-muon* and the *anti-tau* processes are (self) expansions, which were earlier parts of the energy/intensity capacity of the proton and the anti-neutron processes. The intensity impact of the process is less and less, since the expansion works out the intensity capacity (or in our conventional terms: the mass) of the proton and the anti-neutron processes. The expansion is losing on its intensity and speed values. The measured quantum impact is getting less and less.

Collapses *(d-s-b)* in the direct and *(u-c-t)* in the anti-process need drives. They are driven from their fully expanded electron and anti-electron stages to their fully collapsed status, to the *inflexion*, the point with $\Delta t = 0$, where the expansion starts again from. The *inflexions* within the elementary cycles have no time duration. These are is the turning points, repeating the intensity of the collapse into expansion.

There is no difference, which electron process drives hadrons into collapse. The *inflexions* within the proton and the neutron processes happen at the same time moments and the anti-processes happen in parallel. Therefore we will simplify the explanation: the collapse will always mean the neutron process and the expansion the proton process in the followings. The only point to remember is: *the elementary format of the neutron process – in elementary communication – depends on the space-time of the proton process; = on the speed of the quantum communication the electron process drives with*! The proton process is the priority; and the neutron process is the consequence in elementary communication. The question is why?

The answer is coming from the fact, that the proton process has its expansion *(t-c-u)* dominance. The developing extra electron process drive finds the extra *(d-s-b)* in the neutron process and the highest value of the quantum speed prevails. The response will obviously follow, as the *(d-s-b)* line of the "prevailing" impact remains without drive, which will be driven by the (extra) electron process of the other elementary process.

Neutron processes belong to the space-time of the electron process drive!
And the space-time has been established by the anti-proton/proton process inflexion of the elementary process. This is the feeding back from the driven neutron process. The anti-process guarantees the elementary process will be continuing the same way. Until the elementary communication does not change the relations.

The *proton-electron-neutron-anti-neutron-anti-electron-anti-proton* cycle runs between the two *inflexions* and representing all quark/anti-quark relations between the hadrons. The measured (in conventional terms) mass is the measured intensity impact of the change of the internal processes, the actual "energy/intensity source" of the process.

In the case of the *proton* and the *anti-neutron*, and the *electron* and *anti-electron* processes and in the case of all *quark* and *anti-quark* processes in the expansion:

the change of the impact of the self-expansion = the intensity of the work of the acceleration is:	the value of the impact at speed v is:
$\frac{dmc_x^2}{dt_o}\left(1-\sqrt{1-\frac{v^2}{c_x^2}}\right);$	$\frac{dmc_x^2}{dt_x}=\frac{dmc_x^2}{dt_o}\sqrt{1-\frac{v^2}{c_x^2}};$

5B1

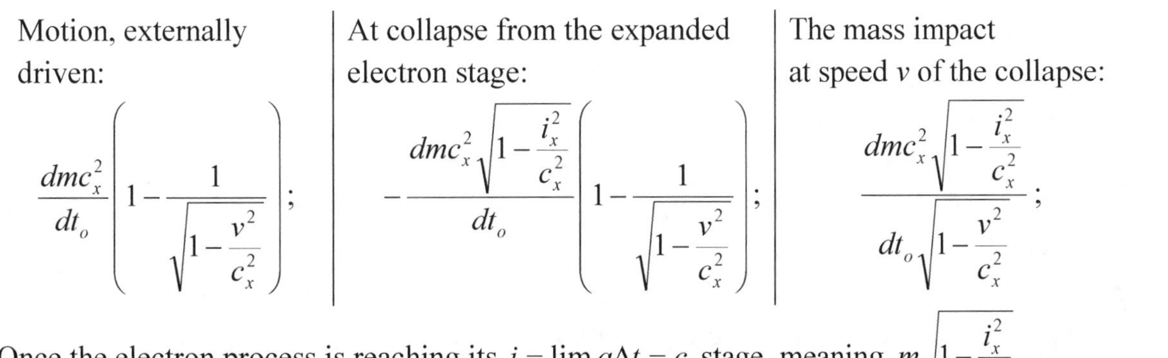

Motion, externally driven:	At collapse from the expanded electron stage:	The mass impact at speed v of the collapse:	
$\frac{dmc_x^2}{dt_o}\left(1-\frac{1}{\sqrt{1-\frac{v^2}{c_x^2}}}\right);$	$-\frac{dmc_x^2\sqrt{1-\frac{i_x^2}{c_x^2}}}{dt_o}\left(1-\frac{1}{\sqrt{1-\frac{v^2}{c_x^2}}}\right);$	$\frac{dmc_x^2\sqrt{1-\frac{i_x^2}{c_x^2}}}{dt_o\sqrt{1-\frac{v^2}{c_x^2}}};$	5B2

Once the electron process is reaching its $i = \lim a\Delta t = c$ stage, meaning $m\sqrt{1-\frac{i_x^2}{c_x^2}}$,

the collapse starts from this stage: The mass, as impact of the collapse returns into its original (minus the entropy product of the cycle) self-expansion full energy starting stage.

The collapse is in fact the slowing down phase of the motion – gathering energy intensity. There are two components here for influencing the mass impact: the time flow, function of the speed and the speed value itself, function of the intensity of the accelerating collapse.

The expansions after the *inflexions* from the neutron and from the anti-proton processes start with the highest intensity capacity, accumulated during the collapse. *Inflexions* happen for *zero* time count, replicate the intensities in the opposite direction.

The mass values of the *proton* and the *neutron* processes represent the highest number of the impacted quantum at the inflexions:

$\frac{dmc_x^2}{dt_o}\beta = \frac{dmc_x^2}{dt_o}\left(1-\frac{(c_x-i_x)^2}{c_x^2}\right);$	with reference to 1B10, the mass of the proton process is formulating at the anti-proton/proton inflexion;	5B3
$\frac{dmc_x^2}{dt_o}\sqrt{1-\frac{(c_x-i_x)^2}{c_x^2}};$	with reference to 1B7, the mass of the neutron process at the neutron/anti-neutron inflexion.	5B4

The difference in the expressions above explains the very small but real difference in the mass impacts of the proton and the neutron processes:
The measured mass of the *proton* is: 938,3 MeV/c^2; of the neutron: 939,6 MeV/c^2;

The expanding acceleration runs out; the intensity of the acceleration is slowing down; in parallel the duration is increasing. The result is quasi constant electron process impact, with quasi constant speed value.

The *inflexion* itself is a passing through the zero speed rest status for zero time,
- ✓ with increased intensity after the inflexion in the proton process;
- ✓ and also with increasing intensity in reaching the neutron collapse.

The *inflexion* means reaching the highest intensity, changing the direction at the top:

The increased intensity impact, which is measured as the W^+, W^- bosons – will be introduced later.

This also explains the increased intensities of the first two quark statuses (*top* and *charm*) in the proton process and the last two quark statuses (*strange* and *bottom*) of the neutron process. And the *anti-top* and the *anti-charm* and the *anti-strange* and the *anti-bottom* statuses in the anti-directions. The *up* and the *down* quark statuses mean approaching speed value $\lim i = c$ of the expansion with increasing Δt of the process, with deceasing acceleration in the process $\lim a = 0$.

There are here three time categories: t_x - the time system of the elementary process; t_o - the time system of absolute rest, which has only its theoretical importance (and practical explanation in Section 1); and dt or Δt - the time system where the impact of the acceleration (the mass) is measured.

The definition of the "mass at rest status" in our conventional understanding is the measured as 938-939 MeV/c^2 - the integrated impact of the intensities of the processes and the quark stages.

The *electron* process is the driving force of the neutron process (all the collapses).

5B5 The starting impact of the electron process corresponds to the end stage of the proton process. $\frac{dmc_x^2}{dt_x}\sqrt{1-\frac{i^2}{c_x^2}}$

The *electron* process is an acceleration at constant speed $i = \lim a\Delta t = c$.

The work of the further acceleration-drive is:

5B6

$$\frac{dmc_x^2}{dt_x\varepsilon_x}\sqrt{1-\frac{i_x^2}{c_x^2}}\left(1-\sqrt{1-\frac{(c_x-i_x)^2}{c_x^2}}\right)=\frac{dmc_x^2}{dt_i\varepsilon_x}\left(1-\sqrt{1-\frac{(c_x-i_x)^2}{c_x^2}}\right);$$

where ε_x is the coefficient, expressing the integrated internal (*t-c-u*)/(*d-s-b*) relation of the proton and the neutron processes. The intensity coefficient of the electron process is the characteristic of the electron process *blue shift* drive of elementary process *X*.

The neutron processes are "neutral", their mass and intensity values correspond to the intensity of the electron process drive of the collapse and, with reference to 5A4, to the relation with the proton process. The proton and the anti-proton processes, as events are similar in all elementary processes, just with the variety of quantum speed and intensity values. (The specifics of these anti-proton, proton events will be explained later.)

Ref. 5B3 5B4

With reference to 5B3 and 5B4, the intensity (mass) impacts of the proton and the neutron processes are almost the same, but still with a small difference. This also explains the increased intensity of the τ and the μ stages of the electron process impacts. The intensities of the *anti-tau* and the *anti-muon* processes, driving the anti-proton process are of the same high value. These intensities correspond to the final stage of the anti-proton collapse.

The intensity of the quantum drive of all anti-electron processes is constant as: $\frac{c_x^2}{\varepsilon_{x-}} = const$

The surplus of the anti-electron process is the quantum impact of *gravitation*.

5.4
Bosons, the quantum impacts
(and about the *weak interactions*)

S.
5.4

There are also events, measured as of very short duration on the *Earth* surface, like *neutrinos* and especially *bosons* including *Higgs*.

The life-times of the W^+, W^-, Z and the *Higgs* bosons are the diapason of

$3 \cdot 10^{-25} - 1.56 \cdot 10^{-22}$ sec – in our time system on the surface of the *Earth*;

and their impact is very significant as for being around the $80 - 126\ GeV/c^2$ region.
So, these bosons, as carriers of the weak interrelation are representing quantum impacts of increased intensity. But there should not be any surprise having these data. Alongside with elementary events of lengthy duration of hadrons and electron processes, we also measure short processes, but of increased intensity.

The estimated duration of the proton process is the magnitude of 10^{29} years and the estimated lifetime of the electron process is of 10^{27} years. At the same time the lifetime of the neutron process has been estimated by the conventional approach as just of minute's magnitude.
We have to note in this case here however, that there is no way to separate from each other the electron process as drive and the neutron process as driven. We also have to note, that the duration of the event-drive and the duration of the event-driven shall be equal with each other. So, the lifetime of the proton, the neutron and the electron processes can be assessed as of the similar magnitude.

For comparison, the intensity of the quantum impacts of the proton, neutron and the anti-proton and anti-neutron processes is around $\approx 0.94\ GeV/c^2$.

This means, *bosons* are extremely high intensity impacts within the elementary processes, acting for extremely short time duration, measured within our space-time on the surface of the *Earth*.

The W^+ and the W^- *bosons* represent the quantum impact of the *inflexions* of the neutron/anti-neutron and the anti-proton/proton processes. This way they well harmonise with our conventional view, as they are the mediators of elementary transmutations of a certain kind indeed – change in the direction of the process.

W bosons are measured as electrically charged. The sign of the charge is coming from the direction of the change:
- *inflexion* from the direct process to the opposite (neutron/anti-neutron) is with "plus";
- *inflexion* from the indirect to the direct process (anti-proton/proton) is with "minus".

The absolute values of their intensities (and the momentum in our conventional terms) are however the same.
The *Z bosons* are the measured conflicts of the quantum impacts between the drives of the elementary processes (the *electron* and the *anti-electron* process) and the cover, the

elementary processes (the *electron* and the *anti-electron* process) and the cover, the intensity potential of the expansion. This conflict is represented by the *neutrinos* and *anti-neutrinos*. This is a kind of "absorption" as our conventional view proposes it: the absorption of the generating energy intensity of the expansion by the energy intensity intake of the collapse. The collapse is driven by the *electron* processes in the direct line of the cycle and by the *anti-electron* processes in the indirect (anti-)line. The transfers themselves have been represented by the quantum impacts, called *neutrino* and the *anti-neutrino* processes.

The energy intensity transfer ensures the parallel run and the energy balance of the expansions (the proton and anti-neutron processes) and the collapse (the neutron and the anti-proton processes). *Neutrinos* and *anti-neutrinos* are neutral; they are the energy intensities of the transfer, initiated by the collapse, driven by electron processes.

The *tau* drive initiates the *tau neutrino* process, with $\nu_\tau = 15\ MeV/c^2$ covering impact, the *muon* drive the *muon neutrino* process, with $\nu_\mu = 0.2\ MeV/c^2$ covering impact and the electron process drive *electron neutrino* process, which is of $\nu_e = 2\ eV/c^2$ covering impact. The *anti-electron* process drive initiates the *anti-neutrino* impact, all with equal $\nu_e = 2\ eV/c^2$ intensity cover, without separate *anti-muon* and *anti-tau neutrinos*, as the intensity of the anti-proton process is equal in all elementary processes.

There is no conflict between the *electron* and the *neutrino* processes and between the *anti-electron* and the *anti-neutrino* processes in normal elementary circumstances. They are vital parts of the same elementary process. The *Z boson* effect is developing in the case of conflict between the drive and the cover.

In the case of the generation of the conflict, the developing *Z boson* is an energy intensity impact, at the level of 91 GeV/c^2 intensity. The intensities of all *neutrinos* and of the conflicting *electron* processes however are the level of 0.120 eV/c^2 - 0.5 MeV/c^2.
The explanation is that

the *electron* process drive and the internal elementary *neutrino* process energy intensity transfer have not just been permanent, without interruption and break and infinite large in their numbers of their impact, but most importantly, what they all do represent is <u>process</u>: propagation of the quantum impacts within the quantum system, rather than certain "flying" particles.
The conflict is resulting in thunderbolt like impact, in intensity, many times of the intensity of a single separated quantum signal.

5C1 $$n\frac{dmc_x^2}{dt_i\varepsilon_x}\left(1-\sqrt{1-\frac{(c_x-i_x)^2}{c_x^2}}\right)=\frac{dmc_x^2}{dt_i\varepsilon_x\sqrt{1-\frac{v^2}{c_x^2}}}\left(1-\sqrt{1-\frac{(c_x-i_x)^2}{c_x^2}}\right);$$ for n electron process impact drives

The summarised intensity of the *neutrinos* at actual v_x speed is the difference of the intensities between the acceleration and the collapse.
The conflict of these impacts results in high intensity *bosons*.

$$\Sigma e_{neutrino} = \frac{dmc_x^2}{dt_p}\left(1-\sqrt{1-\frac{v_x^2}{c_x^2}}\right) - \frac{dmc_x^2}{dt_n}\sqrt{1-\frac{(c_x-i_x)^2}{c_x^2}}\left(\sqrt{1-\frac{v_x^2}{c_x^2}}-1\right) = \tag{5C2}$$

$$= dmc_x^2\left(1-\sqrt{1-\frac{v_x^2}{c_x^2}}\right)\left(\frac{1}{dt_p}+\frac{1}{dt_n}\sqrt{1-\frac{(c_x-i_x)^2}{c_x^2}}\right)$$

All three W^+, W^- and Z *bosons* are certain phenomenon/processes, well representing the energy/mass balance and exchange between the proton/neutron and the anti-neutron/anti-proton processes: the *inflexions* and the intensity cover, transferred by *neutrino* processes.

In the case of the neutron/anti-neutron *inflexion*, the measured difference between the proton *top* as the highest intensity stage of the start of the expansion and the neutron *bottom* as the highest intensity stage of the collapse, the start of the anti-process is:

$$W^+ + Z = e_{top} - e_{bottom} \cong 170\ GeV/c^2 \tag{5C3}$$

and this is also the missing intensity at the anti-proton/proton inflexion between the (*bottom-*) anti-quark at the start of the anti-process and the *top* quark of the start of the proton process: $\left|W^-\right| + Z = \left|e_{bottom-} - e_{top-}\right| \cong 170\ GeV/c^2$. 5C4

These impacts have been measured in certain circumstances as W^+, W^- and Z bosons, as the conventional approach identifies them.

The *Xi-cc boson* represents the progress of the elementary process. The 1.3 GeV/c^2 intensity of the *charm* quark is the measured stage of the expansion of the 170 GeV/c^2 intensity of the *top* quark. The *Xi-cc boson* is the intensity impact of the cover with two *charms* plus *up* quarks. The progress is free and the *t-c-u* composition might turn temporary into *c-c-u*, or *c-u-u*, as the progress dictates; or other intensity values might be measured in the future, representing the elementary progress.

The measured *Higgs boson* may represent another intensity exchange within the elementary process.

5.5
Electromagnetic interrelation

S. 5.5

Electricity is the propagation of the *blue shift* impact of the electron processes. The *blue shift* impacts propagate through the elementary processes of the wire to the other end, without similar electron process potential. Electron process *blue shift* conflict within a wire generates heat. High intensity *blue shift* conflict generates light impact.
The electron process *blue shift* impact propagating within a cable winded in certain direction around a core generates electron process *blue shift* conflict with the elementary

processes of the core. The conflict is forcing the *blue shift* impact of the electron processes of the core in a certain direction. But the *two ends* have their own problems with the flow. There will be electron process *blue shift surplus* at one end and *deficit* on the other. The case is resulting in a classic electromagnetic structure, presented in Figure 5.4.
The elementary process of the core however in not changing, it remains the same.
Electromagnetic interrelation is the effect of the electron/anti-electron processes.

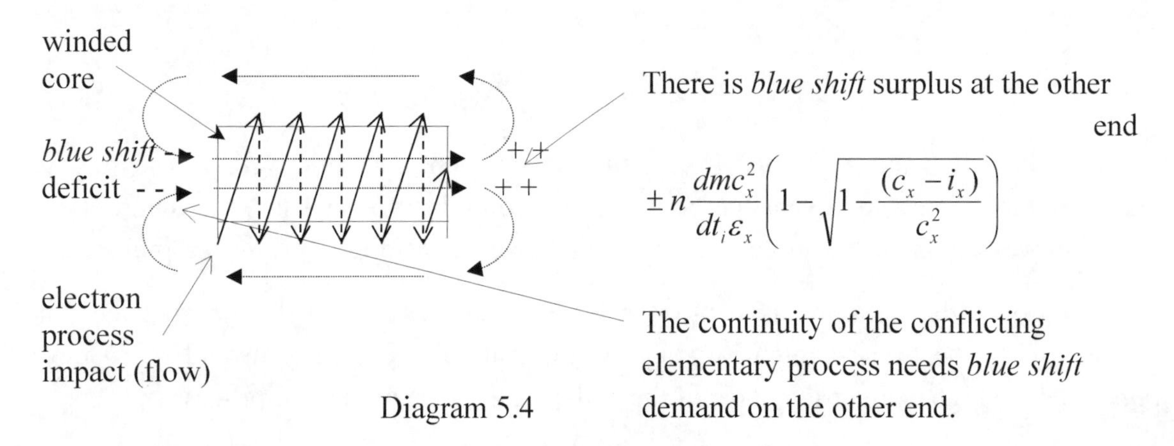

Diag. 5.4

Diagram 5.4

The anti-electron processes, controlling the standards of the elementary cycles of the core, will be reflecting the change in proportions with the power of the electromagnet:

- at the end with the deficit of the electron process impact, the standard *surplus* of the anti-electron process will be *of less* value;
- on the other end, where the electron process is accumulating in surplus, the intensity of the anti-electron process will be of *increased value*.

5D1

The less intensity at one end and the increased intensity of the anti-electron processes on the other end would mean different space-times at the two ends, different than the space-time of the elementary process in the core between.

$$n\frac{dmc_x^2}{dt_i\varepsilon_{x-}}\left(1-\sqrt{1-\frac{(c_x-i_x)^2}{c_x^2}}\right)=$$
$$=\frac{dmc_x^2}{dt_i\varepsilon_{x-}\sqrt{1-\frac{v^2}{c_x^2}}}\left(1-\sqrt{1-\frac{(c_x-i_x)^2}{c_x^2}}\right)$$

The *surplus* at one end results in increased quantum membrane, impacting the intensity of the anti-electron process. The *deficit* on the other end is resulting in anti-electron process impact as well, just in the opposite direction. The function of the anti-side however is the control of the elementary process. Therefore there is a generation of anti-electron process impact within the core in order to correct the elementary process.
In other words: there is no way that the same elementary process in the core would have different space-times with different quantum speed values or there is no way that the elementary structure of the core would be changed.

5D2

The difference in the quantum drives of the anti-electron processes at the two ends of the core is obvious: $\frac{c_x^2}{\varepsilon_{x-}}\neq\frac{c_x^2}{\varepsilon_{x-}\sqrt{1-\frac{v^2}{c_x^2}}}$;

The end of the electromagnet in loss will be looking for the *blue shift* impacts of the anti-electron processes of the other end with surplus in the surrounding quantum space, and adsorbing them in order to restore the standards of the elementary process (the space-time) of the electromagnet at this end.
The quantum space is transferring the quantum impact of the conflict at the end with the surplus. As the "external" quantum space between the two ends is full of other elementary processes, therefore the propagating impact from the end with the surplus will be losing on its intensity on the way in the generating conflicts. This loss is compensated by the intensity capacity of the elementary process within the core.

The propagation of the anti-electron process *blue shift* quantum impact, released by the end of increased intensity might only be limited by the generating conflicts on the way to the other end with deficit.
The overall impact depends on the intensity capacity of the conflict within the core.

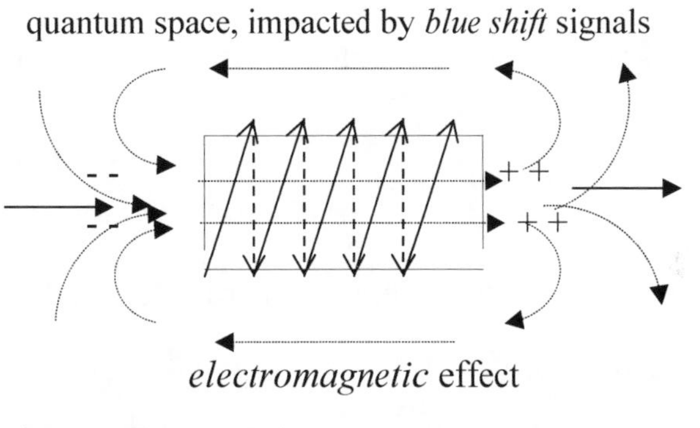

Diagram 5.5

Diag. 5.5

The operation of the ***Earth magnetic field*** is of similar principle.

There is a conflict between the quantum impact of *gravitation* and the *blue shift* impact of the *Sun*. This conflict rotates the *Earth*. As the *Earth* is rotating, all elementary processes above the surface act as a flow of electron processes within the wire around the core. Just here the "wire" is the one at rest and the core is the one which is in rotation.

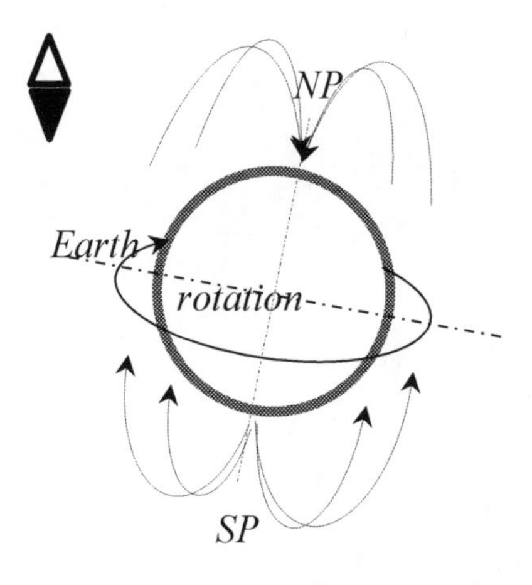

Diagram 5.6

The space above the *Earth* surface is full of the elementary processes of the *Oxygen*, the *Nitrogen*, the *Hydrogen* and the *Helium* processes, all those elementary processes, with less natural quantum speed than the speed of the sphere symmetrical expanding acceleration of the *Earth* surface (the speed of light on our space-time). All these elementary processes are with electron process *blue shift* surplus. In addition, *gravitation* is quantum impact, the anti-electron process *blue shift* quantum impact of the elementary process of the *Earth*.

Diag. 5.6

The acting elementary processes below the *Earth* surface have been impacted. The *blue shift* impacts of the electron processes of the elementary structure of the *Earth* are shifted towards the *South* pole, resulting in surplus in that pole, while the *North* pole remains in demand. For balancing the impact of the rotation, two poles communicate: the quantum

impact at the *South* pole provides the missing anti-electron processes quantum impact to the *North* pole through the quantum space.
Compasses on the surface show this electromagnetic effect in the direction towards *North* pole. This means,

- the space-times of the *Earth* on the *South* and the *North* poles are different; the *North* pole is the one with space-time of increased intensity – (slower time flow)
- the space-time above the *Earth* surface is not just changing, but
- full with electron process quantum impacts.

The intensity of this quantum impact is decreasing with the growth of the altitude above the surface and disappears at a certain height value.

The key of electromagnetic effects is the anti-side, the anti-electron processes.

S. 5.6

5.6 *Gravitational* interrelation

Gravitation is the sphere symmetrical expanding acceleration of the *Earth.*

The *Earth* surface is accelerating under our feet by $g = 9.81\ m/s^2$.

The anti-electron process conflict of the *plasma* is driving the elementary evolution to the *Hydrogen* process, with infinite low quantum speed and intensity values.

5E1

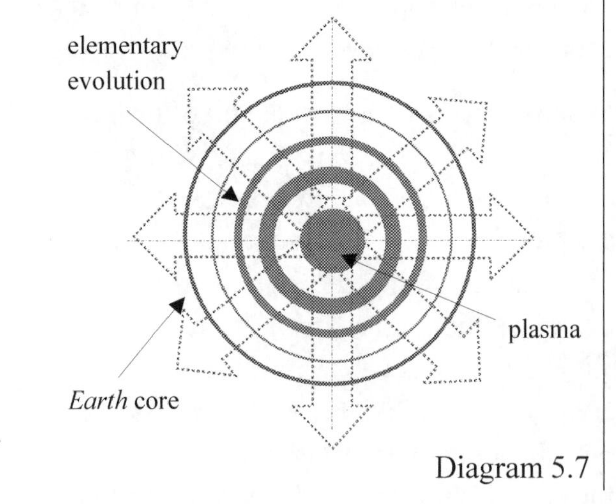

Diag. 5.7

Diagram 5.7

Gravitation – is sphere symmetrical expanding acceleration at constant speed (identical to the electron process)

The speed of the expanding acceleration is:
$\lim i_{Föld} = g \cdot \Delta t = c_{Föld}$

The process on the *Earth* surface is of such a low intensity, that it is unnoticed by us:

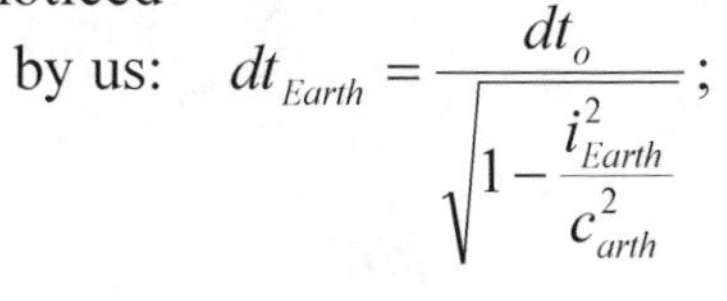

$$dt_{Earth} = \frac{dt_o}{\sqrt{1 - \frac{i^2_{Earth}}{c^2_{arth}}}};$$

$\lim \Delta t_{Earth} = \infty$

5E2

The quantum impact of gravitation is the anti-electron process quantum impact of the elementary processes of the elementary evolution within the *Earth* from the *plasma* to the *Hydrogen* process: $\frac{c_x^2}{\varepsilon_{x-}} = const$

Gravitation has its *mechanical* and *quantum* impacts.

The *mechanical impact* is the sphere symmetrical expanding acceleration of the *Earth* at constant $\lim i_{Earth} = \lim g \cdot \Delta t_{Earth} = c_{Earth}$ speed, a kind of electron process function.

The *quantum impact* is the anti-electron process *blue shift* impact of the elementary processes of the elementary evolution (from the *plasma* state with space-time of infinite high intensity and infinite high quantum speed to the *Hydrogen* process with space-time of infinite low intensity and of infinite low quantum speed). We live on the *Earth* surface between the two ends, within the diapason of the intensity coefficient, value of $0.6 < \varepsilon_{Earth} < 1.1$, with quantum speed value of $c_{Earth} = 299792$ km/sec.

The quantum speed and the intensity values vary, but the anti-electron process quantum drive, the intensity of the quantum impact of *gravitation* is equal for all elementary processes.

Einstein has not examined the case of the acceleration of the *Earth* surface in his theory of general relativity. The *Einstein's* theory proves the *Newton*'s law of universal gravitation. The acceleration of the *Earth* surface however is far not identical to the *Newton's* law on particles attracting each other. In the case of the *Newton's* law, the dominance of the attraction belongs to the mass with higher value and the motion to the other subject of the relation with less mass value. In the case of the supposed acceleration of the *Earth* surface, which in fact should be an obvious condition of any relativistic examination, alongside with the acceleration of any other subjects of the relation, the dominance of the acceleration belongs to the *Earth* surface. (Assessment attached in the Annex.)

Ref. Annex

The relativistic concept requires the examination of all possible options of events.
The acceleration of the *Earth* surface is missing from the official thesis.
The *Pound-Rebka* experiment – the so called practical proof of *Einstein* and *Newton* theories – while it proves the supposed attraction, it also proves the other version, the acceleration of the *Earth* surface.

The *Pound-Rebka blue shift* experiment in simple terms:

- The released down, towards the *Earth* surface, signal is meeting the *Earth* surface, which is expanding and moving upward. The obvious consequence is the increase of the frequency, called *blue shift*.
- If the signal has been released upward, the measured frequency at the level of 22.5 m is of less value (*red shift*), since the expanding upward *Earth* surface is carrying the tower together with the measuring device in the direction of the motion of the released signal.

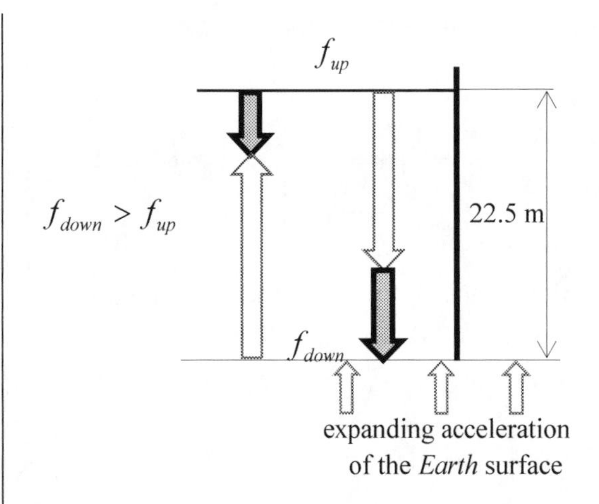

The frequency of the quantum/radio signal released from 22.5 m above the *Earth* surface becomes increased in the conflict with the *Earth* surface!

Diagram 5.8

Diag. 5.8

Let us taking two small balls into our hand. One is from *iron* and the other is form *plastic*. If we let them fall, the two balls reach the *Earth* surface in line with the universal law of gravitation – at the same time moment (for simplicity air resistance influence is excluded).

If we look at the fall of the two balls as events, which are driven down by gravitation, this is the description of the work intensity formula of the gravitation-attraction:

5E3
$$w_x = \frac{dmc_x^2}{dt_E}\left(\frac{1}{\sqrt{1-\frac{(g\Delta t)^2}{c_E^2}}}-1\right);$$

where dt_E is the time count on the *Earth* surface;
c_x is the quantum speed value of the balls, in line with the material of the balls, different for both.
Δt is the time of the acceleration, a $(g\Delta t)$ is the actual speed value of the acceleration.

The equation above reads: the balls are falling down under the attracting impact of *Earth* gravitation. The question is: Is this assumption correct and valid?

For answering this question we have to analyse the formula in 5E3.
It is supposed that Δt is the full duration and Δs is the full distance of the fall. These are equal for the fall of both balls. The work intensities of the attractions are different, since c_x the quantum speed values for the *iron* and the *plastic* balls are different.
Therefore the absolute work values of the fall of the iron and the plastic balls are also different, since $W_x = w_x \cdot \Delta t$ (5E4)

From the relations of the distance and F_x the supposed attracting force is:

5E5
$$W_x = w_x \cdot \Delta t = F_x \cdot \Delta s; \quad \text{and} \quad w_x = F_x \frac{ds}{dt}.$$

Substituting the values in 5E5 into 5E3 and arranging the equation the following way:

5E6
$$\frac{dm}{dt_E}\frac{\Delta t}{\Delta s}\left(\frac{1}{\sqrt{1-\frac{(g\Delta t)^2}{c_E^2}}}-1\right) = \frac{F_x}{c_x^2} = const;$$

The left sides of the equation for both balls are equal, since all parameters on the left hand side are the same for both balls.

In this case however quotient $\frac{F_x}{c_x^2}$ on the right hand side shall also be equal.

Are the quantum speed values for the *iron* and the *plastic* balls different, or not – has no importance. (If different, as the process based approach of this book is proposing, the force values of the attraction are also different.) The main message of the equation however is:

5E7
$$\frac{F_F}{c_F^2} = \frac{F_{pl}}{c_{pl}^2} = \frac{F_x}{c_x^2} = const;$$

the motion under the impact of gravitation toward the *Earth* surface is an *universal event*.

But the equation in 5E6 can be written this way as well:

5E8
$$\frac{dmc_x^2}{dt_E F_x}\left(\frac{1}{\sqrt{1-\frac{(g\Delta t)^2}{c_E^2}}}-1\right) = \frac{ds}{dt} = v_g = const$$

5E8 means, that whatever is the accelerating attraction of the gravitation, **the speed of the motion of the attracted subject in gravitation free fall is constant!**

But the speed values of the subjects in free fall, as experienced are not constant.

The speed of the acceleration relative to the *Earth* surface is $v = g\Delta t$.

What is the meaning of the result in 5E8 in this case?

- since the work intensity formula in 5E3 well specifies the event;
- since the intensity and the full energy of the subject in free fall is increasing.

! This means, the constant speed value in 5E8 **is not about the speed of the subject in free fall**. *This constant speed value characterises another event, the sphere symmetrical expanding acceleration of the Earth, the acceleration of the Earth surface by constant* $v_g = i_{Earth} = \lim g\Delta t = c_{Earth}$ *speed.*

We can justify this in different way as well: Ref. 5E8

If force value F_x together with speed v_g, with reference to 5E8, both are positioned on the right hand side of the equation, that is representing the classical example of relativity, since it describes the increasing and accumulating energy/intensity of the subject in free fall, or the actual work intensity of the acceleration: $w_x = F_x \cdot v_g$ 5E9

$\frac{w_x}{F_x} = v_g$; Dividing this work intensity w_x in 5E9 above by the attracting force value F_x of gravitation in 5E9 also above, it cannot give constant speed value v_g for the subjects in free fall as of 5E8 on the previous page. As the spent work intensities shall result in the acceleration of the iron and plastic balls. 5E10

! Constant speed is only possible if this is about the sphere symmetrical expanding acceleration, the acceleration of the *Earth* surface by speed with $v_g = i_{Earth} = \lim g\Delta t = c_{Earth}$.

In the case, changing our approach and stating that
gravitation is **not about the attraction of the *Earth***, rather
the **sphere symmetrical expanding acceleration of the *Earth* surface**
everything is in order.

We have been accelerated by the *Earth* surface, as *Earth* surface is expanding, carrying us upon. If we drop something from our hand, it will be not accelerating until the *Earth* surface approaches it. Dropped subjects are in “free fall” indeed as they are free from acceleration.

These are not the dropped balls in our example, which are accelerating rather the *Earth* surface is accelerating under the balls. Therefore it is evident that the “gravitational acceleration” of all subjects is identical.

And this is again all about relativity!

How can it be in this case, that I do accelerate in my free fall, while the *Earth* surface is moving by constant speed?

And here the question is not about the $v_g = i_{Earth} = \lim g\Delta t = c_{Earth}$ speed value of the *Earth* surface. The question is, why do I myself and all my subjects accelerate in free fall?

The answer is that at the moment of being free (from any support and acceleration), at the moment the supporting me surface, plate, platform disappears > my speed and the speed of any subjects in free fall becomes immediately decreased, relative to the speed of the *Earth* surface.

! This is not me who is accelerating in free fall. I am in fact slowing down! This is the reason that from my point of view the *Earth* surface is approaching me with acceleration indeed, until the moment the surface approaches me again. And we all know how painful might it be.

The *Earth* surface is the one, which is in expanding acceleration, in fact a kind of electron process, as the acceleration is at constant $\lim i_{Earth} = c_{Earth} = const$ speed value.

This relativistic analysis is missing from the general theory of relativity.

The last and simple question is about the distance we do while accelerating at constant speed together with the *Earth* surface.

The event of the constant acceleration is taken as *v*.
And the relation of the durations is:

$$dt_{Earth} = \frac{dt_{oE}}{\sqrt{1 - \frac{v^2}{c^2_{Earth}}}};$$

v is representing the event,
dt_o is the time count of rest, function of the quantum speed value,
the relation of which is: $c_{Earth} << c_{plazma}$

$$dt_{plazma} = \frac{dt_{opl}}{\sqrt{1 - \frac{v^2}{c^2_{plazma}}}};$$

The result gives infinite high difference in the time count of the event: $\Delta t_{Earth} >> \Delta t_{plasma}$

From the point of view of our space-time, the motion, the event, the acceleration of the *Earth* surface is of infinite low intensity, is of infinite long duration. What is happening with us cannot be sensed in our everyday circumstances for us within our space-time. The low intensity event means low intensity expansion.

S. 5.7

5.7
Photons versus *quantum impulse*

The concern with the definition of the speed of light, as "the speed of photons in empty space" is more than just the controversy of the definition of the *photon*. The main point here is the missing definition of the space. The "empty space" as category is just further deepening the problem. It obviously cannot be equal to "nothing". The definition of *vacuum* and the status of the "almost vacuum" do not improve situation. *Nicola Tesla* made a note in the first decade of the last century, that there is no basis at all stating that the space is curved (is it *Euclidian* or *Newtonian* or other…) without knowing what exactly the space, as such is about.

Looking at the world on process/event basis – meaning, the definition of time cannot be given without events – the answer is coming automatically: *Space has been built up from quantum impulses. Each elementary cycle generate quantum impulse, the entropy product of the cycle. There is no space without elementary processes!*

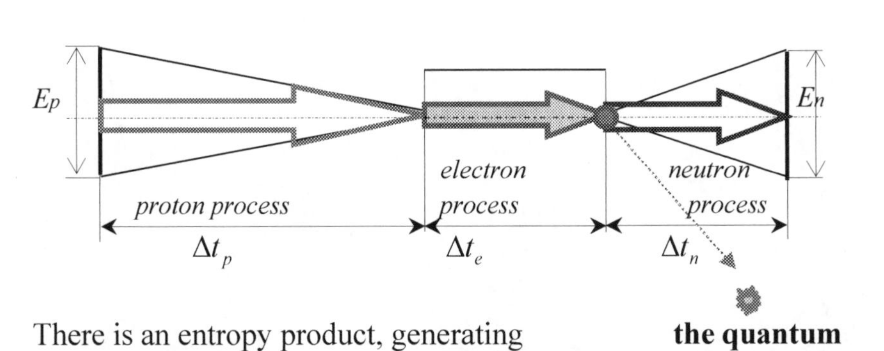

$E_p = E_n$,

$\frac{E_p}{\Delta t_p} \neq \frac{E_n}{\Delta t_n} \quad \varepsilon_p \neq \varepsilon_n$

the intensities of the proton and neutron processes are different

There is an entropy product, generating in each elementary cycle of infinite low intensity **the quantum impulse** = ***the building stones of space***

The entropy products of the intensities of the electron and the anti-electron processes are accumulating and compose the space.

Diagram 5.9

Diag. 5.9

The quantum drives of the elementary process are the electron and the anti-electron processes. Both quantum drives – in their driving function – reach that certain intensity level, when they are not capable to drive the neutron and the anti-proton processes anymore. The driving impact cannot run out to zero intensity and the intensity of the start of the neutron and the anti-proton processes cannot be zero. There is no restart from zero. Running out to zero would result in the completion of the elementary cycle.

So, there is always a remaining intensity potential on the driving side of the cycle. This remaining intensity potential is the *quantum impulse* leaving the elementary process.

Quantum impulses, quantum, quanta are the remains of the electron and the anti-electron process *blue shift* impacts of infinite low intensity. Quantum impulses cannot drive and do not have any driving function. Not just the driving intensity potential has been exhausted, but also the intensity/energy coverage (from the proton process) is missing. Quantum impulses are not in conflict (as they do not have any background intensity reserve and cannot expire to zero); do not take any external quantum impact (as they are in fact *blue shift* impacts themselves, just not capable for any drive). *Quantum impulses* establish and build up the space.

The space, built up from quantum impulses transforms all quantum impacts and signals without any change. *There are no flying photons and there is no empty space*!

Quantum impacts, signals, *information* are propagating within the quantum space composed by quantum impulses, as quantum impulses transfer all impacts without any modification.

All measured experimental findings can be explained on process basis, all of them have its unique and specific function. But most importantly the basis of the explanation is universal for all of them. The basis is the same for the *strong, the weak, the electromagnetic and the gravitation interrelations,* without the sorting them into different categories.

Ch.6

6
Isotopes and their rehabilitation
in line with their space-time and parallel elementary cycles

Isotopes are elementary processes with damaged energy balance and modified space-time. The *proton-electron-neutron-anti-neutron-anti-electron-anti-proton* process sequence has either its energy surplus or energy deficit. The modified space-time is consequence of the balance problems.

Ref. S.3.2

With reference to Section 3.2, alongside with the common time system of the electron and the anti-electron processes, in line with the equal intensities of all anti-electron process quantum drives, the elementary processes have their own quantum speed values.

While all anti-proton processes are driven by equal intensities: $\frac{c_x^2}{\varepsilon_{x-}} = const$;	All anti-proton/proton inflexions and all elementary processes have their own quantum speed values.

6A1

The quantum speed values are different, the time systems of the electron processes are $\lim t_i = \infty$ equal and the electron process quantum drives are certainly different!

The specific intensity value of the electron process *IQ* quantum drive results in different neutron and proton process cycles in line with: $\frac{c_x^2}{\varepsilon_x} \neq const$

6A2

The damage of the elementary balance, the surplus or the deficit of the electron or the anti-electron process drives has their effect on the generating quantum speed value and the space-time of the elementary process.

From all these follows, that

- for *neutron* process dominant elementary processes:

if $\varepsilon_{x(i)} < \varepsilon_x$, meaning, the intensity of the electron process of the isotope is more than the intensity of the standards for the elementary process value,

$\varepsilon_{x(i)-} = \frac{1}{\varepsilon_{x(i)}}$; the anti-electron processes will be in certain increased surplus, as: $\varepsilon_{x(i)-} > \varepsilon_{x-}$.

6A3

if $\varepsilon_{x(i)} > \varepsilon_x$, the surplus of the anti-electron process of the isotope is of less value than the normal.

- for *proton* process dominance the case is the opposite:

with the increase of the electron process surplus $\varepsilon_{x(i)} > \varepsilon_x$ the intensity of the anti-electron process of the isotope is increasing: $\varepsilon_{x(i)-} < \varepsilon_{x-}$.

It is difficult to establish what the primary reason, the surplus at one side or the deficit on the other is. In fact however, this makes no difference.

6A4 The balance problem is impacting the *IQ* value of the anti-electron process and this way either the quantum speed value or the intensity of the anti-electron process: $\Delta n \frac{c_x^2}{\varepsilon_{x-}} = \frac{c_x^2}{\varepsilon_{x-}\frac{1}{\Delta n}} = \frac{c_{x(i)}^2}{\varepsilon_{x-}}$;

As the value of the quantum speed becomes modified, as with reference to 6A1, the isotope status with its internal balance problems is supplemented by the modification of the space-time of the elementary process, which is out of standards.

S. 6.1

6.1
Beta radiation means damage in the electron process

6B1 The intensity of the electron process is the relation of the intensities of the proton and the neutron processes: $\varepsilon_e = \frac{\varepsilon_p}{\varepsilon_n}$;

The damage of the balance either could result in increased or decreased electron process intensity value, relative to the normal elementary status.

Beta-minus radiation means increased electron process intensity and the elementary process tries to give off the *surplus* or work out the *increase* of the electron process.

In the case of *proton process dominance*

6B2

$\varepsilon_e > 1$ $\frac{d\varepsilon_e}{dt_i} > 1; \longrightarrow \frac{d\varepsilon_{e-}}{dt_i} < 1;$ With the external utilisation of the surplus the elementary process becomes balanced again. The space-time of the process becomes normal.	The electron process surplus will be released: the elementary process finds other elementary processes for electron process impact. The release of the intensity surplus reduces the intensity of the quantum membrane of the anti-electron process: with the decrease of the intensity of the anti-electron process the anti-proton/proton inflexion will be of less intensity.

This damage is equivalent to increased proton process dominance.

In the case of *neutron process dominance*

6B3

$\varepsilon_e < 1$ $\frac{d\varepsilon_e}{dt_i} < 1; \longrightarrow \frac{d\varepsilon_{e-}}{dt_i} > 1;$ The external utilisation means the use of number of parallel elementary cycles. And this way, the increased intensity of the electron process returns to its standards.	The increased intensity of the electron process means increased surplus of the anti-electron processes on the anti-side. The isotope utilises its own electron process drives of increased intensity through elementary communication. By the utilisation of the intensity increase, the extra surplus of the anti-electron process disappears. The external utilisation of the electron process may cause damage to other elementary processes.

This damage is equivalent to increased neutron process dominance.

The external use of the parallel elementary cycles, means, that some of the acting parallel cycles, increased in relative numbers as result of the damage, will drive neutron processes outside the isotope. As the proton process cover will be coming from the elementary process of the isotope, this in fact results in the damage of the effected elementary cycle.

Beta-plus radiation is consequence of the decrease of the intensity of the electron process. The isotope tries to use the electron processes quantum impact from aside, from other elementary processes.

In the case of *proton process dominance*

$\varepsilon_e > 1$ $\frac{d\varepsilon_e}{dt_i} < 1; \longrightarrow \frac{d\varepsilon_{e-}}{dt_i} > 1;$ With the use of external drives the number of the neutron processes becomes less and the elementary process is capable for internal drive. This is the way out of the damage!	The surplus of the electron process is decreasing. It results in the decrease of the intensity of the *anti-electron process*. With the decrease of the intensity of the *anti-electron* process the drive of the intensity of the anti-proton process becomes even less. The isotope tries to utilize external drives for the neutron processes. The external drive however takes away the elementary cycle.

6B4

In the case of *neutron process dominance*

$\varepsilon_e < 1$ $\frac{d\varepsilon_e}{dt_i} > 1; \longrightarrow \frac{d\varepsilon_{e-}}{dt_i} < 1;$ The external electron process, driving the neutron process "takes away" cycles and this way the demand in load in the direct line becomes less, corresponding to the intensity norm of the elementary process.	The decreasing intensity of the electron process is result of the decreasing surplus of the *anti-electron process*. The anti-electron process is still in surplus, but the intensity of the process does not correspond to the norms of the elementary process. There is a need for external support for corresponding to the standards.

6B5

The self-rehabilitation of these kinds of isotopes is based on the existence and the operation of parallel elementary cycles. The change in the number of the elementary cycles demonstrates the damage. Isotope in fact means the variety of the atomic weight of the elementary process at equal periodic number. The communications of the isotopes with the environment is the tool for impacting the parallel cycles, as explained, and re-establish the internal balance. This is however not about standard elementary communication neither in proton nor in neutron process dominant cases. The external elementary processes are "seduced" by the *isotopes*, the damaged elementary processes.
Elementary processes however are always ready for help and communication in any form of it.

There are no flying *beta* particles in the case of *beta* damage. *Beta* damage in both cases is the deviation within the quantum communication of the elementary process. The *blue shift* impact of the electron processes is either too intensive or does not correspond to the intensity need of the elementary process.

S. 6.2

6.2
Alpha and *gamma* radiation

Alpha radiation is the heavier format of the *beta-minus* isotope status. For re-establishing the standard space-time of the elementary process, the intensity surplus of the process is given off by the release of the proton process.

6B6

$\varepsilon_{x(isotope)} < \varepsilon_x$; as $\varepsilon_{n(isotope)} > \varepsilon_n$; Releasing proton process intensity however is only possible as part of the complete elementary cycle. The smallest completed cycle belongs to the *Helium* process. *Helium* processes are released. $$\frac{c_x^2}{\varepsilon_{x(isotope)}} = \frac{c_x^2}{\varepsilon_x} + \frac{c_{He}^2}{\varepsilon_{eH}}$$	Neither the release of the electron process nor the release (from the increased surplus) of the anti-electron process helps; Therefore there is no other way, just releasing proton processes as direct consequence. The preconditions of the release in the case of proton and neutron process dominance are different, but the consequences are identical. The reason of the increased intensity of the proton process is the increased intensity of the *anti-proton/proton* inflexion.

Because of the released complete elementary process, *alpha* radiation is the only one in fact with quasi "mechanical" impact. This is the impact of the released complete *Helium* process in quantum communication.

Gamma radiation is the even more serious format of the *beta-minus* radiation. The *gamma* radiation is usually the escorting impact of the *beta-minus*.
Gamma radiation is the direct release of the anti-electron process surplus. This is the reason *gamma* radiation is with ionising impact. Anti-electron processes of equal *IQ* drive directly impact other elementary processes.
The point here is the increased value of the quantum speed. The internal control of the elementary process keeps the anti-electron quantum drive at the necessary level, but the increased quantum speed modifies the space-time of the elementary process. The release of anti-electron processes is necessary.

6B7

$\varepsilon_e > \varepsilon_{e(isotope)}$; $\varepsilon_{e-} < \varepsilon_{e-(isotope)}$ $$c_{x(isotope)}^2 \cdot \varepsilon_{xe(isotope)} = \frac{c_{x(isotope)}^2}{\varepsilon_{xe-(isotope)}} = const$$ For keeping the space-time of the elementary process in order, anti-electron processes are released.	With the increase of the intensity of the electron process, the surplus of the anti-electron process is increasing. In order to compensate the increase of the intensity of the anti-electron process, anti-electron process impacts are released.

The released anti-electron processes are with damaging impacts.
The modified quantum speed values, increased or decreased, may modify the space-time of the elementary processes, resulting in damages.

6.3
X-ray and *neutron* radiation

S.
6.3

X-ray radiation is result of the quantum impact of the *blue shift* conflict of the electron processes of certain of elementary processes.
There is a generation of electron process *blue shift* conflict if certain elementary processes, like *Fe, Ni, Co, Cu* are bombarded by electron process *blue shift* of limited intensity. The conflict generates electron process *blue shift* impact, which can be used externally as *X-ray* radiation for different purposes.

Neutron radiation is no other than the consequence of the destruction of elementary processes under the quantum impact of increased electron process *blue shift* conflict of the nuclear fission. The destruction is generating elementary processes with heavily damaged elementary balance. The generating elementary processes with damaged structure impact the environment by *beta*, *alpha* and *gamma* radiation.

There are no flying neutrons in the neutron radiation. Neutron radiation means the energy/intensity damage of the elementary process of the isotope – demand for proton process intensity cover – by the elementary processes of the damage.

6.4
Rehabilitation of isotopes

S.
6.4

The intensity of the electron process of the isotopes in the case of *beta-minus*, *alpha* and *gamma* is higher than the standard normal. The increased intensity of the electron process means, the intensity of the neutron process is higher than the intensity in its balanced elementary status. As the intensity of the drive of the standard process is of less value, the intensity increase means conflict in the communication with other elementary processes.

The conflict, generated by the increased intensity of the electron process of the isotope is the radiation impact itself. The intensity of this impact depends on the actual format of the radiation:

- the easiest is the *beta-minus*, as this is a “simple” electron process *blue shift* impact;
- while *alpha* is rare format, its damaging impact is a heavy one, since this means a complete elementary process;
- *gamma* is an escorting impact of the *beta-minus* – as the function of the anti-process is always to ensure the internal balance; *gamma* is acting alone only in the case, if all other chances for self-correction has already been expired.

The higher the periodic number of the isotope is, the higher is the intensity impact. And as the intensities of the electron processes of the elementary processes with high periodic number are significantly higher than the intensities of the elementary processes of our life processes, the conflicting (radiating) impact might be in this case especially high and damaging for us.
The clarification of the reasons of the *isotope* status, discussed earlier gives the chance for speeding up the rehabilitation of the damaged status of the elementary processes.

Isotopes can be subjects to external, deliberate quantum impacts.
External electron process *blue shift* impact from other elementary processes generates conflict within the isotope. The result of the conflict is heat generation. In the case of *beta-minus*, *alpha* and *gamma* this conflict works out the damage (the intensity unbalance): the generating energy/intensity surplus leaves the elementary process is the form of heat.

In the case of missing intensity capacity, (in the case of *beta-plus*) the conflict may contribute to the return back of the elementary standards. The acting in the damaged elementary process less intensity invites other elementary processes to communicate with the damaged process. In this certain case the communicating process is the one, which is losing on its intensity – for the benefit of the isotope.

The question is, what the elementary composition of the communicating external process shall be?
The communicating impact should not just be an elementary process accepting, taking the damaging impact of the isotope. It must be a one, causing real conflict within the elementary process of the isotope!

Conflicts can be generated if the communicating elementary systems have their elementary impact on the isotope. Normal elementary processes would contribute to the utilisation of the increased intensity of the isotope, but will not speed up the rehabilitation. The conflict needs an elementary mix in equilibrium or close to equilibrium status, still with slight electron process surplus. In this case the *blue shift* impact of the electron process with increased intensity of the isotope will be reflected back to the isotope. This way the isotope is meeting in fact itself within the conflict. The conflict generates heat and the balance is back.

Ref. Diag. 3.2

Elementary cycles mean the running in parallel elementary processes.
With reference to Diagram 3.2, certain selection from the first 29 elementary processes of the Periodic Table gives the chance to compose a mix with close to quasi balanced status. The advantage of this kind of a mix is having the necessary "reflecting back" feature, which helps to create the conflict within the isotope.

The equilibrium status of the mix makes the composition strong enough for withstanding the *blue shift* impact of the isotope. Instead the electron process *blue shift* impact of the mix is the one, having effect on the isotope. Reflecting back the *blue shift* impact of the isotope, the mix generates conflict and the isotope loses its increased and damaging energy potential in the heat generation of the conflict, initiated in fact by its own.

The components of the mix include almost all elementary processes of the Periodic Table up to the periodic number 29. *Cuprum* process in various mineral formats. Oxides, carbonates, nitrates, silicates, calcites and sulphates can be used in certain proportions with the minerals of the elementary processes listed in Diagram 3.4 on the next page. *Carbon*, *Nitrogen*, *Oxygen*, *Magnesium*, *Silicon*, *Sulphur* and *Calcium* processes in fact do not change the number of their cycles. The change in the *Sodium*, *Aluminium*, *Potassium*,

Chlorine and *Phosphorus* processes is minimal. The selection from the *Titanium – Cuprum* elementary line is necessary for increasing the overall intensity level of the electron processes of the mixture.

Elementary process	PN	Atomic weight	ε_{e-}	ε_e	c_x	IQ_x
Titanium	22	47.90000	1.1610	0.86133	325403	1.229E+11
Vanadium	23	50.94150	1.1983	0.83452	330589	1.310E+11
Chromium	24	51.99600	1.1503	0.86934	323901	1.207E+11
Cobalt	27	58.93320	1.1664	0.85734	326159	1.241E+11
Cuprum	29	63.54000	1.1746	0.85135	327304	1.258E+11

Table 6.1

Table 6.1

The consistency of the mix of the minerals with these elementary processes is powder-like. Adding the decontaminating mix to the isotope and mixing it well it generates heat and results in immediate rehabilitation. Mixing it with water the result is solid concrete structure of increased hardness and water tightness. The hardening process of the concrete structure is escorted by the heat generation of the conflict of the elementary communication. Water is with electron process surplus and proton process potential. The principle of the formulation of the isotopes with *beta-minus*, *alpha* and *gamma* radiation are similar, just the consequences are different.

Elementary process	PN	Atomic weight	ε_{e-}	ε_e	c_x	IQ_x
Lithium	3	6.94	1.2961	0.77155	343815	1.532E+11
Beryllium	4	9.012	1.2362	0.80893	336777	1.394E+11
Boron	5	10.81	1.1458	0.87275	323299	1.197E+11

Table 6.2

Table 6.2

Because of the relative high internal intensities and the low number of the parallel elementary cycles *Lithium*, *Beryllium* and *Boron* processes can take in the intensity impact of elementary conflicts. This is the reason *Boron* process is used as controlling and slowing down process of the chain reaction in nuclear reactors.
The chain reaction in nuclear reactors can be stopped if the electron process conflict, resulting in the fission of the *Uranium* processes – the nuclear fuel – can be slowed down or stopped. The same "swallowing the intensity" characterises the *Cadmium* process, just of higher periodic number and atomic weight.

Cadmium	48	112.410	1.3244	0.75506	347549	1.600E+11
Lead	82	207.800	1.5153	0.65994	371754	2.094E+11

Table 6.3

Table 6.3

Starting from the *Cadmium* elementary process all following elementary processes in the Periodic Table with increasing atomic weight have a kind of shielding function. The intensity of the electron process drive of these elementary processes is quite increased for reflecting back *beta*, *alpha* and *gamma* impacts.

Ch.7

7
The *space* and the *space-time*

The *quantum impulse*, the *energy quantum* is establishing the space!
Elementary processes from the *plasma* to the *Hydrogen* process exist and operate within the space, which has been established by the quantum impulse of their own elementary processes. While there are definite differences in the quantum speed and the intensity values of the elementary processes, the generating *quantum impulses* have equally infinite low value of *IQ* quantum drive: $\lim e_{quantum-impulse} = 0$

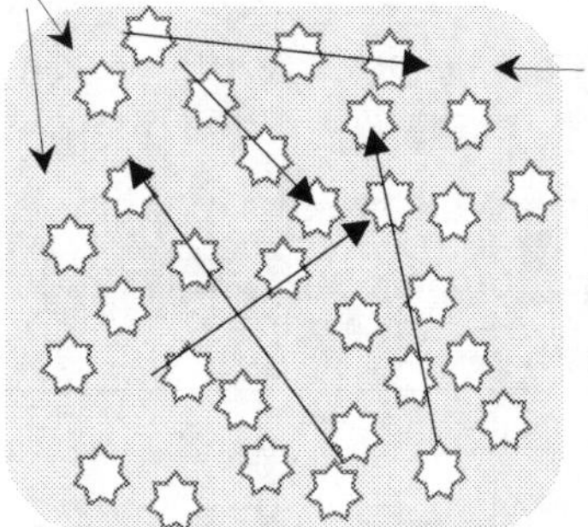

there is no space without event (without elementary process)!

The quantum space is established by elementary processes.

Fig. 7.1

Figure 7.1

The generating quantum impulses cannot disappear, cannot turn into "nothing" and they are not capable to any drive function, as

(1) they are out of elementary processes, and
(2) they are the existing smallest intensity impacts.

Quantum impulses, *quantum*, *quanta* are transforming all quantum signals in the quantum space without any modification. The speed of the transfer depends on the quantum speed of the quantum signal.
Elementary cycles have their start and end statuses. While the actual elementary cycle replaces the previous one, the generating *quantum impulse* does not disappear with the renewal of the cycle. *Quantum impulses, quantum, quanta* are accumulating.

The space, quantum space is the quantum impulse product of the elementary processes.
The quantum impulse does not have any driving intensity capacity to move away and there is no impact which would move the generating impulses away from the elementary cycles. Therefore, quantum impulse in fact belongs to the elementary process of its generation. This way the internal quantum system (space) generated and established by elementary processes is in constant extension through the boundaries of the elementary processes. It occupies more and more "space", but the points of the extension are the proximate regions of the elementary processes.

As the anti-proton/proton processes, as events, are similar for all elementary cycles and as the neutron processes are neutral and driven – the key factor of the quantum impulse generation is the IQ drive of the electron process.

$$c_x^2 \cdot \varepsilon_x = \frac{c_x^2}{\varepsilon_{x-}} = const \qquad \text{7A1}$$

While the anti-proton processes have been driven by quantum drives of equal intensities and while the proton processes, all are similar, the number of the impacted quantum and quantum speed values are different. The number of the cycles has been determined by the intensity of the (controlling) anti-electron processes.

The quantum drive of the anti-direction of the *plasma* corresponds to the standard value of the anti-proton process drive of the elementary evolution: $IQ_{pl-} = IQ_{x-} = const$

$$IQ_{pl-} = \frac{c_{pl}^2}{\varepsilon_{pl-}}; \quad \Bigg| \quad \lim c_{pl} = \infty, \text{ and } \quad \lim \varepsilon_{pl-} = \infty \qquad \text{7A2}$$

As the intensity of the anti-neutron expansion is repeating the intensity of the neutron collapse and the quantum drive of the *Hydrogen* collapse is infinite high intensity, as

$$\lim IQ_{pl} = \lim \frac{c_{pl}^2}{\varepsilon_{pl}} = \infty; \quad \Bigg| \quad \lim c_{pl} = \infty, \text{ and } \quad \lim \varepsilon_{pl} = 0 \qquad \text{7A3}$$

With reference to 3F7 and 3F8: Ref 3F7 3F8

$$n_{n+1} = n_n\left(1 - \frac{\varepsilon_{x-} - 1}{\varepsilon_{x-}}\right) = n_x \cdot \varepsilon_x;$$

for neutron dominance, if $\varepsilon_{x-} > 1$; and

$$n_{n+1} = n_n\left(1 - \frac{\varepsilon_x - 1}{\varepsilon_x}\right) = n_x \cdot \varepsilon_{x-};$$

for proton process dominance, if $\varepsilon_x > 1$

The number of the elementary cycles of the collapsing – by infinite high quantum speed and intensity – *Hydrogen* processes, the initiation of the plasma was: $\lim n_H = \infty$.

This means, the $\lim n_{pl1} = \infty$.

For neutron process dominance:

- the *higher* the intensity of the electron process is, the higher the numerical value of ε_{x-}, the intensity coefficient of the anti-electron process is, the higher is the intensity of the neutron/anti-neutron inflexion, the *higher* is the surplus of the anti-electron processes, the higher is the number of the operating in the elementary process parallel cycles; the *higher* is *the gradient* of the decrease of the number of the operating elementary cycles (the less is the value of ε_x is).
  - the higher the number of the operating elementary cycles is, the higher is the surplus (the number) of the anti-electron processes, the intensity of which expires as quantum impact of gravitation.

For proton process dominance:

- the *higher* the intensity of the electron process is, the less the numerical value of ε_x the intensity coefficient of the neutron/anti-neutron inflexion is, the *higher* is the number of the elementary cycles, the *higher* is the electron process surplus and the *higher* is *the gradient* of the decrease of the number of the operating elementary cycles;

- the *higher* the number of the operating cycles is, the higher is the number of the electron processes utilised by the elementary process in elementary communication.

There are here two important points to be noted here:

1. The *Hydrogen* process does not utilise its electron process surplus, as there is no operating cycles, which have been completed;
2. The *plasma* does not have operating cycles, as the initiation comes from the *Hydrogen* process.

The two together compose an elementary cycle!

The data of the elementary evolution strengthen each other and compose the whole. All proton processes, as energy source of the evolution, with the variety of quantum speed and intensity values, are similar and products of equal anti-electron process *IQ* drives.
As the generation of the quantum impulse is proportional to the operating elementary cycles, from the above follows that

- the less the intensity of the generation of the quantum impact of gravitation is, the higher is the relative number of the generation of the quantum impulses; and
- the higher the intensity of the generation of the quantum impact of gravitation is, the less is the relative number of the generation of the quantum impulse.

Ref. Table 3.1

With reference to Table 3.1, the balance of the quantum communication and the intensity relation of the elementary processes are guaranteed: Elementary processes with higher cycle numbers and with higher *IQ* values are of increased cycle reduction gradient. The number of the cycles of the elementary processes with proton process dominance is less, but they keep their cycle numbers at quasi stable level. The number of the elementary cycles is important, since quantum impulses are generating at the end of each completed cycle. Either as result of the electron or the anti-electron process drives.

We have to distinguish the meaning of the *quantum space* in general, established by quantum impulses from the meaning of the *space-time*.
The definition of *space-time* differs from the definition of the *space: space-time* is the segment of the *space*, where events happen by certain quantum speed and intensity values.
The specificity of the elementary world is that space is established by elementary processes (events) and these events themselves are part of the space.
Space-times vary, as trees vary in gardens. The garden itself is the space and all trees within the garden represent a certain space-time. While each of the trees needs certain approach and brings different fruit, each of them are vital parts of the garden.

Space-time means a distinct quantum system within the space with specific quantum speed and intensity values and certain time count. Space-times belong to elementary processes.

The space-time of each elementary process has its definite quantum speed and electron process intensity value. Therefore identical events happen for different time counts in each of the elementary processes. Identical events might happen in parallel in elementary processes, but for different internal time counts, durations.

The question is what identical events in elementary processes are for?
The answer is easy, as the proton process is the only one, with energy/intensity source. All other processes are consequences. The anti-neutron process is a certain energy source as well, but it is the consequence of the neutron collapse and part of the self-control.

Ref. 1A1 1A3

Elementary processes generate quantum impacts, which are transferred by the quantum impulses of the space. The quantum impacts of space-times with higher quantum speed and intensity approach larger distance (more number of quantum impulses) for the unit period of time.

Plasma is a process, with infinite high quantum speed, with infinite high intensity of the electron process, with infinite high number of cycles and of infinite large space-time (consequence of the quantum speed of infinite high value).
The generation of the *plasma* is initiated by the *Hydrogen* process:

> The accumulating electron process of infinite low intensity and the accumulating proton cover capacity of the *Hydrogen* process is completing in the *plasma* process.

The space for the *plasma* process has been established by the generating *quantum impulses* of the immediate completion of the neutron collapse of all *Hydrogen* processes. The infinite high quantum speed of the *plasma* ensures the quantum communication is of infinite wide "distance" (subject to later explanation) in any direction and in fact contains the space-times of all other elementary processes.

S. 7.1

7.1
The space

The time system of the *plasma* is infinite "short" as the quantum speed and the intensity of the process are of infinite high values. The intensity of the quantum drive is of infinite high value and the intensity coefficient is of infinite low value.
The cooling of the *plasma* process means the step by step increase of the time system, originally infinite short at the *plasma* state. This cooling process is the elementary evolution itself. The cooling also means the decreasing tendency of the anti-electron process surplus, the quantum impact of *gravitation*.

The *plasma* is losing on the original intensity of its generation (the result of the electron/neutron collapse of the *Hydrogen* process). *Plasma* represents the elementary *global turn-around inflexion* and the start of the new cycle of the elementary evolution. Plasma is a *process* of infinite high intensity and of infinite slow time count and of infinite large space-time.

Each step of the cooling of the plasma means a different time system. Looking at the plasma from the time system of the elementary processes (the results of the cooling process) the "originally" infinite slow time system of the plasma is speeding up. From the point of view of the elementary processes, the lifetime of the plasma is infinite long with infinite long quantum impact of gravitation. (With reference to the equal intensities of the anti-electron processes, the duration of the effect of the quantum impact of gravitation – belonging to the quantum speed of infinite high value of the plasma – is infinite long.) The elementary processes themselves are the results of the cooling process. The farer the elementary process from the plasma status within the line of the elementary evolution is, the smaller is the space-time of the process and longer is the experienced duration of the plasma process projected within.

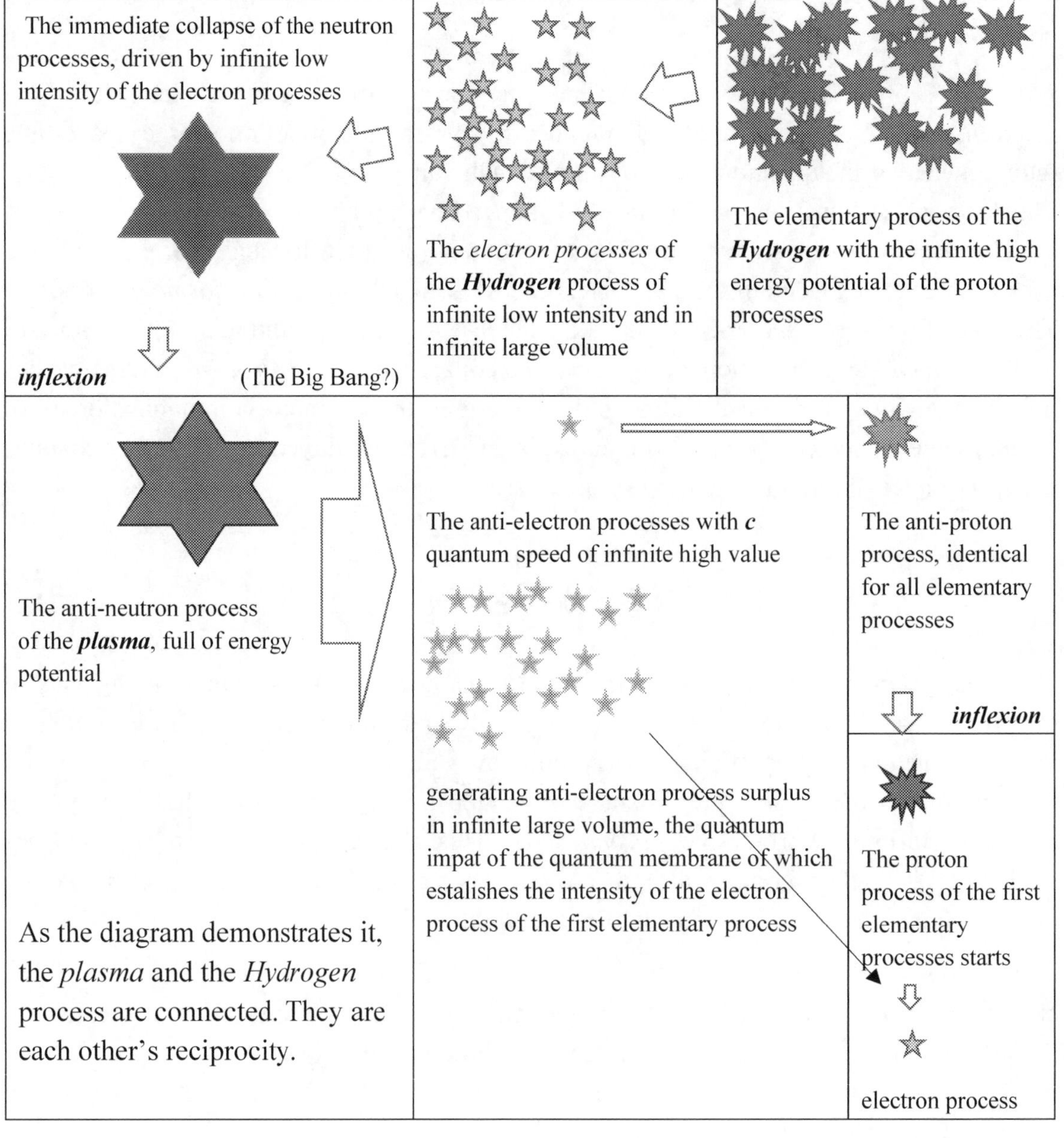

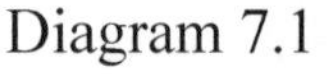

Diag. 7.1

Diagram 7.1

This is the case within our time system on the surface of the *Earth* as well. Our impression is that the *plasma* status as a process is of infinite long duration. The plasma status however is a process of infinite high intensity – which means in fact infinite short process time within the time system and the space-time of the plasma process. The infinite short process time in its own time system, the infinite large space-time and the infinite long effect of the anti-electron processes are the literally related components of the plasma and the elementary evolution.

All elementary processes, consequences of the cooling process of the plasma are happening in parallel. As result of the cooling, the plasma is losing on its "turn-around-inflexion-for-infinite-short-time" status and step by step turns into sub-plasma, sub-sub-plasma and elementary process statuses, including ourselves on the *Earth* surface, ending up with the *Hydrogen* process, completing the global cycle of the elementary evolution.
In absolute terms this is obviously one and the same cooling process. In intensity terms however the process has its intensity scale of infinite variety, with its decreasing tendency from the plasma to the *Hydrogen* process. The definition of the *plasma* state covers high number of intensity statuses, which can also be interpreted as sub and sub-sub *plasma* statuses. The elementary processes of the Periodic Table represent the specific stages of the cooling process. The plasma goes through infinite variety of intensity and quantum speed values up to the identification of the elementary processes within the core of the *Earth*.
It is difficult to establish where the plasma status as such ends and where the "real" elementary process phase starts, but the known elementary processes start at the level of the intensity coefficient of the electron process of $\varepsilon_{start} > 0.634$ and end at around $\varepsilon_{end} < 1.015$ within the scale of $0 < \lim \varepsilon < \propto$.

- The plasma and the sub-plasma statuses mean "elementary processes" with infinite large number of elementary cycles with infinite high electron process intensity and with infinite number of anti-electron processes, acting for infinite long time period.
- The *global turn-around Hydrogen/plasma inflexion* happens; the evolution is progressing; the conflict of the infinite number of elementary processes acting in parallel remains. This in fact represents the real status of the plasma.
- The time system of all anti-electron processes (and also of the electron processes) is identical, but the quantum speed values and the *IQ* quantum drives are different.
- The space-time of the plasma is determined by infinite high quantum speed and intensity values of the inflexion. With reference to Section 1, this means infinite high intensity quantum drive and infinite slow time count. With the progress of the elementary evolution this integrated quantum drive is decreasing.
- The anti-electron process based elementary control drives the elementary evolution between the two ends: from the *plasma* to the *Hydrogen* process. The quantum speed and the intensity of the processes of the elementary evolution are decreasing step by step. The proton processes, as events and starting points are similar, but with different quantum speed and intensity values for all elementary processes.

S. 7.2

7.2
The distance, quantum impacts of different space-times make

Each elementary process has its own space-time. The quantum speed and the intensity values characterise the elementary process.

The space-times of elementary processes communicate within the universal quantum system, space. The key of the quantum communication is dt_i the equal time count for all electron processes. Quantum space transfers quantum signals in line with the *IQ* drives of the space-times of the elementary processes of the origin of the impact. Space-times have their own quantum speed values, as one of their most important characteristics and have the intensity of their quantum signals, which is based on the intensity of the electron process.

- the quantum speed is the basis of the internal time count,
- the intensity of the electron process represents the "length" the quantum signal can impact the space.

7B1 $$\frac{dmc_x^2}{dt_i\varepsilon_x}\left(1-\sqrt{1-\frac{(c_x-i_x)^2}{c_x^2}}\right);$$

The *unit time period of the change*, within the intensity formula of the electron process *blue shift* impact as signal, is: $dt_x = dt_i\varepsilon_x$

7B2 The quantum speed value of the elementary process however shall correspond to the relation of the distance the quantum impact makes and of the unit period of time of the change: $c_x = \frac{ds_x}{dt_x}$; and this way: $ds_x = c_x \cdot dt_x$

What we call internal distance and understand as the measurement of the space is the number of quantum impulses the quantum signal approaches for the unit period of time.

7B3 With reference to Section 1.7 of "*The Quantum Impulse and the Space-Time Matrix*" the higher the quantum speed is, the smaller is the *internal distance* of the space-time: $ds_x = \frac{c_x^2 \cdot \varepsilon_x}{c_x} dt_i = \frac{C}{c_x}$;

The *internal distance* is function of a single parameter: the quantum speed value.

C means constant in 7B3: $C = const$, as $c_x^2 \cdot \varepsilon_x = const$ and the dt_i time systems of all electron processes are identical.

7B4 At the same time, the higher the intensity of the quantum drive is, the larger is the space-time, the higher is the number of the quantum of the space-time: $IQ = \frac{c_x^2}{\varepsilon_x}$;

The higher the value of the quantum speed is the smaller is the internal world for the given space-time.

If the value of the internal distance $\lim s_x = 0$ is, the internal distance is infinite short, the space-time itself is infinite "large": the signals, generating within this space-time are approaching infinite large number of quantum impulses. (Like plasma.)

If the value is $\lim s_x = \infty$, the internal distance is infinite large and the space-time is infinite "small", as the signals, generating within this space-time are approaching infinite small number of quantum impulses for the unit period of time. (Like the *Hydrogen* process.)

Size parameters "large" and "small" are definitions of relativistic meaning. Relativistic, since the number of the quantum impulses, approached by quantum impacts always depend on the duration of the event as well. An event within a relatively "small" space-time, acting for long time count may mean the same event acting within a "large" space-time for short period of time. Time and space have been connected. The comparison is only possible if either the time measurements or the numbers of the impacted quantum are identical. (Otherwise the difference cannot be detected.

Example:

The duration of the plasma process, as an event, impacting certain number of quantum is of infinite short duration within its infinite large space-time, while the same plasma process event, with the same number of impacted quantum gives infinite long duration within our space-time on the *Earth* surface.

The event is the same, the durations are clearly different:

$s_{pl} = f(n_{pl}) = \Delta t_{pl} \cdot c_{pl}$; versus $s_{pl} = f(n_{pl}) = \Delta t_E \cdot c_E$; result in: $\Delta t_E = \Delta t_{pl} \frac{c_{pl}}{c_E}$; 7B4

The internal distance of the space-time of the plasma process and of our space-time on the surface of the *Earth* is different.

The quantum communication of the plasma space-time is of infinite high intensity. Its space-time is infinite large. The quantum communication on the surface of the *Earth* is characterised by the quantum speed of our light signal. Our space-time has its definite limits.

The *internal distance* and the *size in general* are relativistic categories, but these both characterise the efficiency of the quantum communication. The shorter the value of the *internal distance* is, the higher is the efficiency of the quantum communication of the given space-time – the shorter it takes to the quantum signal to propagate through the space in time. The quantum signals of less efficient space-times make the same distance (approaching quantum impulses) for longer time flow.

We can characterise space-times as small or large, but it shall also be noted, that the distance a quantum signal makes – in other words: the number of the quantum impulses the signal is approaching – whatever the size of the space-time is, always is the function of the intensity reserve, the capacity of the impact, acting, specifying the duration of the impact.

With reference to 7B1, 7B2, 7B3 and 7B4, there is no way the quantum speed value of the quantum signal of an elementary process is slowing down. If the quantum signal has the sufficient reserve within the space-time of the elementary process, the signal exists. If not, the signal disappears! (This is not about the step by step contraction of the space in line with the slowdown of the quantum speed.)

Ref. 7B1 7B2 7B3 7B4

$ds_y = \frac{c_y^2 \cdot \varepsilon_y}{c_y} dt_i = \frac{C}{c_y}$; and $c_y < c_x$;	There might be quantum signals with different quantum speed values, different than the original one, but these must be different signals of different elementary processes, with different space-times!

7B5

dt_i the time system of the elementary communication is one and the same for all elementary processes, the characteristics of space-times however are clearly different.

Quantum signals of higher intensity and higher quantum speed values may impact other space-times of less quantum speed values of the elementary processes; they might generate conflicts with others, but they either do exist as the quantum speed of certain space-time or elementary process or do initiates the next step of the elementary evolution.

Events are one and the same for all space-times within the overall space, while the time counts are different. Space-times either communicate, or are in conflict with each other.

Quantum impulses (quantum) are of equal and infinite low intensity for all space-times.

In the case of infinite large number of events and of infinite high number of quantum impacts, the overall space cannot be capable equally transmit the quantum impacts of all events. This is the case when conflicts are generating with the increase of the intensity values.

Conflicts are vital parts of the propagation of quantum impacts and the communication of space-times. Quantum signals can only propagate within the space through different space-times, if the intensity of the *IQ* quantum drive of the propagating signal prevails comparing with the *IQ* value of other space-times on the way of their propagation. This consequently also means that the propagating quantum signals lose on their intensity/energy values.

The space-time of the plasma originally is of infinite high size, but the intensity of its quantum impact is losing on its value going through all other space-times, the products of the elementary evolution. The original intensity is decreasing and finally results in the space-time on the *Earth* surface, equal to the quantum impact of *gravitation*.

! ! ! The quantum impact of *gravitation* and the *Earth* surface space-time prove the harmony of the elementary evolution. The quantum impacts of all elementary processes are losing in the conflicts with the quantum impacts of other space-times, coming through from the centre of the plasma to the *Earth* surface, including the *Earth* core with the known elementary processes, ending up with equal quantum impacts which correspond to the quantum impact of the space-time on the *Earth* surface and equal to the quantum impact of *gravitation*.

The fact that the reduction of the intensities of the quantum impacts and the quantum speed values of the space-times of the elementary evolution result in intensity value, equal to the quantum speed value and the quantum impact of gravitation is the *necessary/sufficient condition* of the establishment of the space-time of the *Earth* surface.

The intensity of the quantum impact approaching the *Earth* surface from below and above cannot be different. The heat generation of the plasma and of the subsequent statuses of the plasma prove the conflict of the propagation of the quantum impacts of the internal elementary processes. Those elementary processes which have already passed the quantum speed value of the space-time on the *Earth* surface in their elementary evolution are in the permanent conflict of the gaseous status.

7.3 Communication of space-times

S. 7.3

The communication of elementary processes in fact is the communication of space-times. Space-times are with specific ruling quantum speed and intensity values. The tools of the elementary communication are the electron process *blue shift* impact, the neutron process to be driven and the proton process, the intensity cover of the *blue shift* impact.
The basis of the communication is the communication between the elementary cycles of elementary processes. This is communication between space-times.

Different space-times are communicating if the electron process quantum impact of one of the elementary processes is driving the neutron process of the other elementary process. Classical quantum communication means the reciprocal drive and proton process cover of each other's neutron processes. This communication needs elementary process with increased intensity of the electron process at one side and elementary process with electron process surplus on the other. The elementary process with electron process *blue shift* surplus is initiating the communication, but the other elementary process, with increased electron process intensity is the one, dominant of the communication.

The communication is difficult if the space-times are positioned at certain distance from each other. The communication at distance, with reference to the previous section, means the quantum impulses transfer the quantum signal, which might be meeting conflicts on the way approaching the space-time of the communication. The generating conflict either stops the communication, if the conflicting impact is of higher *IQ* quantum drive value, or simply reduces the intensity of the propagating impact. Therefore the space-times of elementary processes have their distance limits in communication.

The best example for elementary communication at distance is the communication of living organism with the surface regions of the *Earth*; with minerals, with the sand cover of waterfronts; with hills and with the gases of the atmosphere.
Overcoming the distance in these cases is easy.
Quantum signals can also be generated as result of electron process *blue shift* conflicts and sent away as technical products of the conflict of elementary space-times.

7C1

a/
$$n\frac{dmc_1^2}{dt_i\varepsilon_1} = \frac{dmc_1^2}{dt_i\varepsilon_1\sqrt{1-\frac{v^2}{c_1^2}}} = \frac{dmc_1^2}{dt_i\varepsilon_2};$$

a./ The <u>*valid option*</u>: the conflict shortens the acting time of the electron process, this way increases the intensity of the impact.

b/
$$n\frac{dmc_1^2}{dt_i\varepsilon_1} = \frac{dm}{dt_i\varepsilon_x}\frac{c_1^2}{\sqrt{1-\frac{v^2}{c_1^2}}} = \frac{dmc_2^2}{dt_i\varepsilon_1};$$

b./ The <u>*false option*</u>: the change of the quantum speed would mean different space-time.

In the valid option a./ the conflict, we may say, the intensity value of the electron process is the one increasing, while its space-time remains unchanged.

This way $dt_x = dt_i \cdot \varepsilon_x$, the acting time of the impact shortens, increasing by that the *IQ* drive of the electron process.

The increase of the intensity is well presenting the case, as the energy/intensity generation of the elementary process becomes increased at the standard quantum speed value, without any change of the space-time.

7C2 The conflict means: $\varepsilon_x^{conflict} < \varepsilon_x^{st}$; and $\varepsilon_{x-}^{conflict} > \varepsilon_{x-}^{st}$; but $\frac{c_x^2}{\varepsilon_{x-}} = c_x^2 \cdot \varepsilon_x = const$;

The increase in the value of the intensity coefficient of the anti-electron process on the anti-side increases the anti-electron process surplus. This surplus on the anti-side results in the increase of the intensity of the quantum membrane of the electron process at the direct side (the intensity coefficient of the electron process becomes of less value): the electron process drive of the collapse of the neutron process becomes of increased intensity.
The lower intensity of the anti-electron process (the coefficient is higher in its value) would stop the increasing tendency of the elementary process and would manage it back to normal, but the external impact is in effect. And the conflict is re-generating again and again until the external influence stops.

It shall be noted that all generating conflicts have their limits on the elementary process. There is no way elementary processes would tolerate it without the destruction of their structure if the conflicting impact is overdosed.

The example for the increased *blue shift* impact of a conflict is the one, generated by us in our information technology in selected elementary process space-times. The generated electron process quantum impact propagates in our space-time. The intensity of the *IQ* quantum drive of the impact is higher than the acting quantum drive in our space-time. The signal goes through until the increased intensity at the space-time of the origin provides the necessary backup capacity.

The other example is the generation of light: The conflict is generating within the space-time of the elementary process of the lightening device whatever the elementary process, meaning metal or gas, of the conflict is. The external, from our electricity system supplied electron process *blue shift* impact is the one, which is initiating the conflict within the lightening device. (The intensity level of the electricity supply from our power plants shall correspond to such a high level, which allows feeding the conflict, going through all parts of the electricity system.) The quantum impact of the increased conflict propagates through our space-time and lights it up, until the capacity of the conflict allows it. Closer to the lamp the density of the impact is higher, with the increase of the distance the impact of the conflict is less and less. Once the supply is off, the conflict expires, the light effect is over. The permanent conflict sooner or later destroys the elementary process of the device.
The light impact is product of a technical impact. The conflict is impacting our space-time and propagates within our space-time.

Space-times slow down the propagation of the quantum impacts, results of conflicts. Quantum impacts lose on their intensity and *IQ* drive until its value becomes insufficient to go through the elementary conflicts of the space-time of the propagation. This is not about the loss of the kinetic energy of the quantum signals, separated in their motion from the origin of their generation.

$$n_x \frac{dmc_x^2}{dt_i \varepsilon_x}\left(1-\sqrt{1-\frac{(c_x-i_x)^2}{c_x^2}}\right)-\Delta e_{loss} < \frac{dmc_y^2}{dt_i \varepsilon_y}\left(1-\sqrt{1-\frac{(c_y-i_y)^2}{c_y^2}}\right); \qquad 7C3$$

This is exactly about the energy intensity of the origin. Once the intensity of the conflict at the space-time of the origin, reduced by the intensity of the losses is less than the intensity of the space-time of the propagation, the impact stops, the signal disappears.

7.4 Computer communication

S. 7.4

When we use the *Silicon* process, with significant electron process *blue shift* surplus for saving and storing our data in our computers, the basics of the quantum communication are the parallel cycles of the *Silicon* elementary process. We are the ones, who modify the elementary cycles of the *Silicon* process by the impact of our data in the variety of the capacity of our device. The impacts are so tine that they can modify only the cycles of the elementary process in surplus. The *Silicon* process, having electron process surplus is ideal for the function. There is infinite high number of elementary cycles in the silicon process and the impacted ones operate as modified.
Letter "*s*" saved, for example will always be letter "*s*", or our videos, saved will always be as they were saved, until the modified cycles come to their ends. As the experienced duration of the proton processes in our space-time is about billion years, there will be no problem with the accuracy of the saved data.

The universal space contains infinite number of space-times. Signals propagate from space-times through space-times in the space. Signals of less quantum drive intensity cannot impact space-times with higher quantum drive intensity. This is the reason we can observe space-times of higher quantum drive intensities: The signals of our quantum communication sent are reflecting back by the *blue shift* impacts at the boundaries of the space-times with the higher quantum drive intensities. All elementary processes with higher quantum drive intensities than the one on the *Earth* surface can be seen. All others, the *Hydrogen, Helium, Oxygen, Nitrogen* processes cannot.

This is also the reason external quantum impacts (spies and feeds) can approach our computers, in fact whatever protection we apply against it. The infinite variety of the modification of the *Silicon* elementary process can always give the intensity of the signal necessary to find us.

This is quantum communication, based on the generation of the conflict of the electron processes. If the intensity of the external *IQ* quantum drive is similar to the one of the certain data in our computer and if the capacity of the spies is high enough to find the resonance with them, the communication is established. This in fact is good enough for reading our thoughts.

The impact of the conflict, the change in one of the space-times (elementary process) may impact the quasi similar other space-time at distance. The definition of "quasi similar" means the space-time, the status of which corresponds to the changed status of the original space-time.
What is the reason?
Similar space-times, even separated and distanced from each other represent the same system, the same space-time, as their quantum speed values are equal.
Let us see the example:
There is an electron process conflict (elementary or technical one, no difference) within a space-time, a change, generating a quantum impact:

7D1 $$\Delta e_{conflA} = n\frac{dmc_A^2}{dt_i\varepsilon_x}\left(1-\sqrt{1-\frac{(c_A-i_A)^2}{c_A^2}}\right)$$; later we will only refer to the *IQ* value

7D2 The generation of the *IQ* value of the quantum signal becomes also different: $$IQ_{confA} = n\frac{c_A^2}{\varepsilon_x} = \frac{c_A^2}{\frac{\varepsilon_x}{n}} = \frac{c_A^2}{\varepsilon_y} = IQ_B$$

The change generates a new IQ_B drive with c_A, ε_y values.
The conflict, generated in space-time *A* in 7D1 will be impacting similar space-time(s), with the corresponding quantum speed and IQ_B quantum drive. The direct communication is possible, as the resonance with the impacted data is without conflict.

The first condition of the communication is the necessary intensity capacity reserve of the quantum impact of the communicating "spy" space-time to overcome the *blue shift* conflicts on the way to the targeted space-time.
Calculating the loss at the distance of the communication, the arriving quantum signal will be corresponding to the similar space-time at the end of the receipt with c_A and ε_y data.

7D3 The acting impact drive on the similar space-time is: $$\frac{c_A^2}{\varepsilon_y} = (n-n_{way})\frac{c_A^2}{\varepsilon_A}$$

There is here again the earlier important aspect to be noted: Quantum signals are propagating if the necessary intensity capacity source at the origin of the impact is guaranteed. Quantum signals do stop immediately if the intensity of the feeding source disappears. Quantum signals are not like rain drops leaving the cloud and coming down separated. If the energy supply of the origin disappears, the signal stops. There is no time delay in the stop.

7.5 Source of energy generation

S. 7.5

Conflicting space-times are source of energy generation, if the communication is selected the right way.
Constant effect of quantum impacts may generate constant conflict. Constant conflict is an energy source within the space-time, receiving the impact. With the external release of the quantum impact of the conflict, the permanent generation of the conflict is an energy intensity resource.

The natural conflict within the *pyramid* is generating a certain increased internal *IQ* drive. The pyramid has its certain elementary composition and space-time; the permanent quantum impact of gravitation generates conflict within the pyramid. If we prepare a similar space-time, with similar quantum drive, the two space-times can communicate on their similar elementary basis. The communication in this case may result in the taking away, pumping out the generating within the pyramid *energy* to the similar space-time.
The only limit is the capacity of the space-time of the energy generation (*pyramid*) for overcoming all conflict of the supply on the way of the propagation of the energy impact.

The principle is simple: Once the connection and the resonance have been established, the two space-times are representing one and the same space-time. In the case of any balance deviation at one end, the other end compensates it. If any conflict (energy) is taken away from our, for this purpose prepared, space-time the natural source of the pyramid will compensate it and provides the necessary missing impact (energy). Similarly to the continuity of the water supply from deep down to up, if the pumping force has been remained in effect. The only challenge is to create the necessary space-time for establishing the quantum communication.

Ref. 8C8 8C11 of QS S-TM 7D3

With reference to 8C8 and 8C11 of Section 8 of *"The Quantum Impulse and the Space-Time Matrix"*, the *IQ* drive of the conflict within the *Great Pyramid of Giza* is:

$$IQ_{pyr} = 1.46739\text{E}+11 \; km^2/\sec^2 \, .$$

This quantum drive value, at the aggregate electron process intensity of the pyramid of $\varepsilon_{pyr} = 1.00421$ corresponds to the equivalent speed value of the quantum communication: $c_{pyr} = 383{,}870$ km/sec.

The pyramid has its internal/external absolute energy balance. But the internal conflict within the pyramid is of high intensity and a space-time for communication as well.

The internal conflict within the pyramid generates that certain quantum speed value which ensures the quantum impact of gravitation on its external surfaces and provides at the same time an energy intensity capacity for use.

Ref. S. 4.3.1

With reference to the experiment in Section 4.3.1 and Figure 4.7, the two small scale pyramids are communicating.

S. 7.6

7.6
The space is one and the same and space-times contain each other

The space-time of the highest quantum speed is containing all other space-times.
The space-time of the *plasma* contains all other space-times of the elementary processes.

There is no way for having similar space-times at distance from each other!
Similar space-times mean the same space-time. The space itself is the accumulation of quantum impulses. And the distinguishing parameter of the space-times is the value of the quantum speed of the communication.

What does in this case "*distance*" in its conventional understanding on event basis mean?
It means the number of the impacted quantum.
While the number of the impacted quantum can be the same, the impacting signals might be different. In this case we say: we do an equal distance for different time counts, which is function of our speed value of quantum communication.

Fig. 7.2

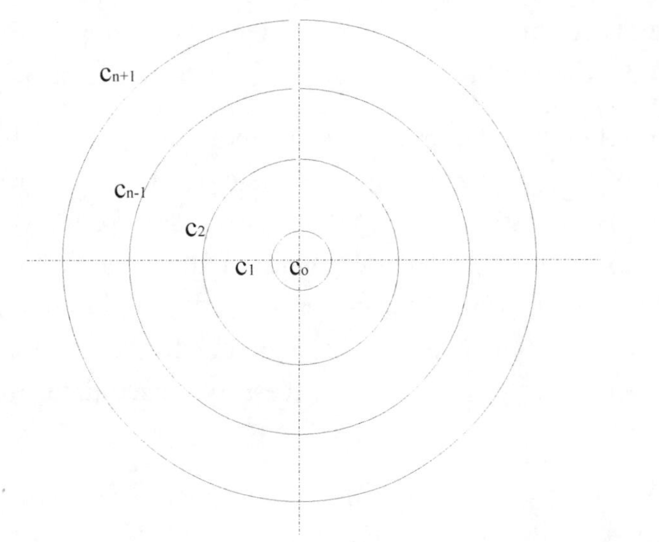

The event is one and the same, happening within all existing space-times, containing each other.

The duration of the event depends on the space-time of the measurement!

Figure 7.2

The meaning of the space-time exactly is this: communication by different quantum speed values.
Can we say that similar space-times are "located in space at a distance from each other"?
$c_x^2 \cdot \varepsilon_x = const$ and

7E1 $$dt_x = \frac{dn_x}{c_x}; \quad \frac{dn_x}{dt_x} \cdot \varepsilon_x = const; \quad \frac{dn_x}{dt_i \varepsilon_x} \cdot \varepsilon_x = const;$$

7E2 We are coming to a new formulation of the equality: $\frac{dn_x}{dt_i} = const$;

It means: the intensity of the approach of the quantum signals in any quantum speed values are equal. In other meaning: the number of the quantum impulses approached by the electron process *blue shift* impacts of elementary processes, for the unit period of time, is equal. This makes quantum communication possible.

The different is in the intensities of the quantum impacts!
Impacting the same quantum number by higher quantum speed means higher intensity, as

$$dt_x = \frac{dn_x}{c_x}. \quad dt_i\varepsilon_x = \frac{dn_x}{c_x}; \quad \varepsilon_x = \frac{dn_x}{dt_i c_x}; \text{ as } dt_i = const \text{ indeed.}$$ 7E3

Two similar space-times of equal quantum speed, but “distanced” from each other compose one and the same space-time, as the quantum signal for “making the distance” does not need any additional intensity. Quantum impulses do not mean any loss on the way of the communication. Any change in the intensity, therefore will be compensated by the space-time at the “distance”. The $c_x^2 \cdot \varepsilon_x = const$ relation cannot be lost.

The existence of other space-times in the quantum space obviously has its certain influence “on the way of the communication of the *similar* space-times”. The propagation of the own signals of space-times – in conventional terms – might be impacted by other existing space-times. [“Own signal” means actually that certain quantum impact – the event – which is generating the space-time of that certain quantum speed and intensity value (the elementary process).]
But the terminology “similar space-times”, earlier used was only for the explanation in conventional way. They are in fact *one and the same space-time*.

$\varepsilon_x = \frac{dn_x}{dt_i c_x}$ means: The highest the quantum speed of the quantum impact, generated by the space-time is, the highest is the intensity of the quantum impact of the space-time. 7E4

This principle, represented by our elementary world is: space-times have been built on each other and in each other:
The space-time of the plasma (with reference to Figure 7.2) of the highest quantum speed and intensity contains all other space-times, ending by the space-time of the *Hydrogen* process of infinite quantum speed and intensity. Ref. Fig. 7.2

The equal impact of gravitation is result of the effect of the impacts of all space-times on each other. The space is generated by the elementary processes and the highest space-time contains all other space-times. The highest loss of the space-time communication of “the own quantum impacts” belongs to the space-time of the highest quantum speed and intensity, as this is the one containing all other space-times.
The space is one and the same.
The space-time with the lowest intensity and with the lowest quantum speed value is the space-time of the *Hydrogen* process, which in fact not capable impacting the space-time of any other elementary process.

The main message of the discussion is, that

- space-times contains each other;
- the variety of elementary processes means the variety of space-times; but
- similar elementary processes represent the same space-time – in conventional way:

- similar events are of the same space-time;
- space-times are taking care of the balance of their existence, whatever is the "distance" between the events of their generation;
- as space-times have been built into each other and on each other, the assurance of the balance needs the use of additional intensities, results in losses;
- the losses of the natural communication of space-times might be preventing the real communication, but
- this does not change the simple principle: similar events (elementary processes) represent the same space-time!

8
The *matter*

The *matter* is *quantum impact.*
The space is the *matrix* of *quantum impulses*, the generating by elementary processes quantum impacts of infinite low intensity, *quantum, quanta.*

All starts by the collapse of infinite high intensity, the collapse of the accumulated electron/neutron processes of the *Hydrogen* process into *plasma.*
The accumulating *Hydrogen* processes, the final stage of the elementary evolution, with all proton process potential of the evolution, with infinite number of electron process *blue shift* drives of infinite low intensity result in immediate neutron process collapses in infinite numbers. The immediate collapse of infinite high intensity, for infinite short time results in *plasma* status, with the transformation of all accumulated proton process potential into the *plasma.*

The immediate collapse and the *plasma* is about elementary electron process *blue shift* conflict of infinite high intensity: $n_{pl}\dfrac{c_{pl}^2}{\varepsilon_{pl}} = \dfrac{c_{pl}^2}{\dfrac{\varepsilon_{pl}}{n_{pl}}}$; 8A1

$\lim n_{pl} = \infty$ - the number of the acting *blue shift* impacts is infinite high

The *plasma* state means infinite high number of electron process *blue shift* impacts in effect, resulting in a conflict of infinite high intensity. The conflict is the *plasma* itself. The quantum speed of the communication is infinite high; the time count of the event is infinite short within a space-time, size of infinite large.
The status of the matter is *plasma*, with quantum speed value of $\lim c_{pl} = \infty$. 8A2

The plasma conflict within the centre is a kind of extra dense status of the quantum impact, result of the infinite high intensity of the neutron collapse, with the infinite high conflict of the generating anti-electron process *blue shift* impacts.
The decrease of the intensity and the slowdown of the quantum speed of the conflict mean the cooling of the plasma: the less dense conflicting, liquid-gaseous fluid lava and the hardening stage, the *Earth* surface status of the plasma.
The cooling of the conflict of the *plasma* is the appearance of the elementary processes.

The cooling of plasma means the decrease of the intensity of the acting anti-electron process conflict: the plasma is giving off its intensity. As result of the cooling, the intensity of the conflict is decreasing, the intensity of the acting repelling quantum impacts becomes less and less. The cooling process is creating the surface regions of the plasma with less quantum speed value. The conflict of the anti-electron process *blue shift* impacts is decreasing towards the surface.

The cooling is identical to the gradual decrease of the speed of quantum communication, the gradual decrease of the intensity of the anti-electron process surplus and conflict and the corresponding gradual decrease of the acting space-times. The hardening of the surface is the conflict with quantum speed values of our well known elementary processes. Our solid subjects are the matrix of quantum impacts of the elementary processes with quantum speed value and with intensity, quasi without conflict – the solidified status of the original conflicting stage of the *plasma*.
The *matter* is an event, the *matrix* of acting intensities, result of quantum impacts.

The conflicting high frequency intensity status is of high temperature, infinite large quantum speed, and infinite large space-time with infinite short time count.
The "normal" status of the matter is results of the quantum communication of elementary processes at the diapason of the "normal for us on the *Earth* surface" value of the speed of quantum communication, with increased process time – a *matrix* of elementary processes in solidified, liquid and gaseous aggregate states.

The *matter* is not about particles!
The *matter* is the matrix of events in time, result of the quantum communication, the quantum impact of elementary processes. The *matter is the matrix of information.*

The *matter* at the *plasma* state represents a *conflict*, a collapse and neutron/anti-neutron *inflexion* of infinite high intensity; generating anti-electron process surplus in infinite high volume and parallel elementary cycles in infinite high numbers, with electron process *blue shift* drive of infinite high intensity; keeping the intensity of the anti-proton/proton *inflexion* at infinite high quantum speed and intensity values; while the intensity of the generating proton processes, with the definition of certain, but for each elementary process different numbers of the impacted quantum decreasing, with the time count increasing.

There is an infinite low intensity quantum impact without any conflict at one end, called *Hydrogen* process, and another one with infinite high intensity and conflict on the other, called *plasma*. There is an infinite high proton process potential of infinite low intensity, without use, at one end, and the same potential, just in action with infinite high intensity on the other. Quantum impact with infinite low intensity at one end and quantum impact of infinite high intensity on the other.
The *plasma* and the *Hydrogen* process statuses belong to each other. The direct and the indirect sides of the same cycle, connected by the elementary evolution.
The cooling of the *plasma*, the loss of intensity creates elementary statuses. The infinite high intensities are transforming into elementary processes of less intensities, with specific quantum speed and intensity values. The transformation creates lava, gaseous, liquid and solid aggregate states and generates *gravitation*. But the quantum impact, as the origin of the *Hydrogen/plasma* status remains the basic component of all statuses of the elementary evolution – the *matter*.

Holy Bible reads in Genesis 3.19: *"for dust thou art and unto dust shalt thou return"*. A right formulation.

8.1
Once again about the *matter*

S. 8.1

Because of the anti-electron process *blue shift* conflict of infinite high intensity, the *plasma* state is easy to visualise. The visualisation of all other elementary processes is difficult.

The infinite high density of the *blue shift* impacts of the anti-electron processes, their repelling effect of infinite high intensity and frequency create a conflicting, "more than the effect of fire" – is the status of *plasma.*

The cooling of the plasma means the natural decrease of the conflicting status. The cooling process of the plasma has its complexity, as the conflict is decreasing in line with the decrease of the space-time of the process. The generation of anti-electron process *blue shift* impacts – *gravitation* – decreases the acting intensity towards the surface of the conflict. The loss on the intensity of the conflict is equivalent to the decrease of the temperature. The quantum speed and the intensity values of the elementary events are decreasing. The conflict is less and less and the "dense plasma" status turns into different "plasma-lava" statuses. These changes are the symptoms of the decreasing tendency of the conflict, the decreasing values of the quantum speed and the intensity of elementary processes. The events of the proton processes remain similar, just the aggregate statuses of the elementary process vary in line with the cooling process, the change of the intensities.

There will be a stage when, as result of the cooling effect of the self-expansion of the plasma, the direct impact of the plasma conflict disappears, leaving the quantum impact of gravitation acting: we find elementary processes communicating with each other – in order to find the best balanced status for each of them – in the solid status of minerals.

The cooling of the plasma is natural need. Elementary processes and natural events try to find balanced communication with the environment. There cannot be a plasma status existing alone in a quantum space, with external environment without conflict. The self-expansion means the gradual decrease of the intensity, in conventional terms giving off the surplus of energy.

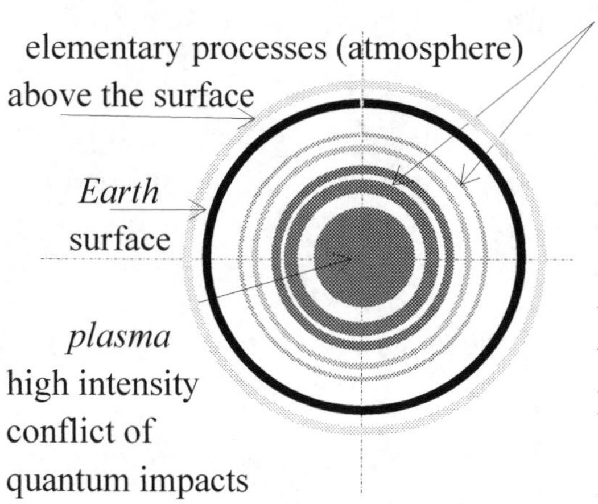

elementary processes (atmosphere) above the surface

Earth surface

plasma high intensity conflict of quantum impacts

Cooling the plasma results in different aggregate statuses of the conflict of the quantum impacts.

The elementary processes of the periodic table have intensity coefficient from *0.6* to *1.1.* All other statuses of the matter are between the infinite high intensity of the *plasma* and the infinite low intensity of the *Hydrogen* processes.

Diagram 8.1

Diag. 8.1

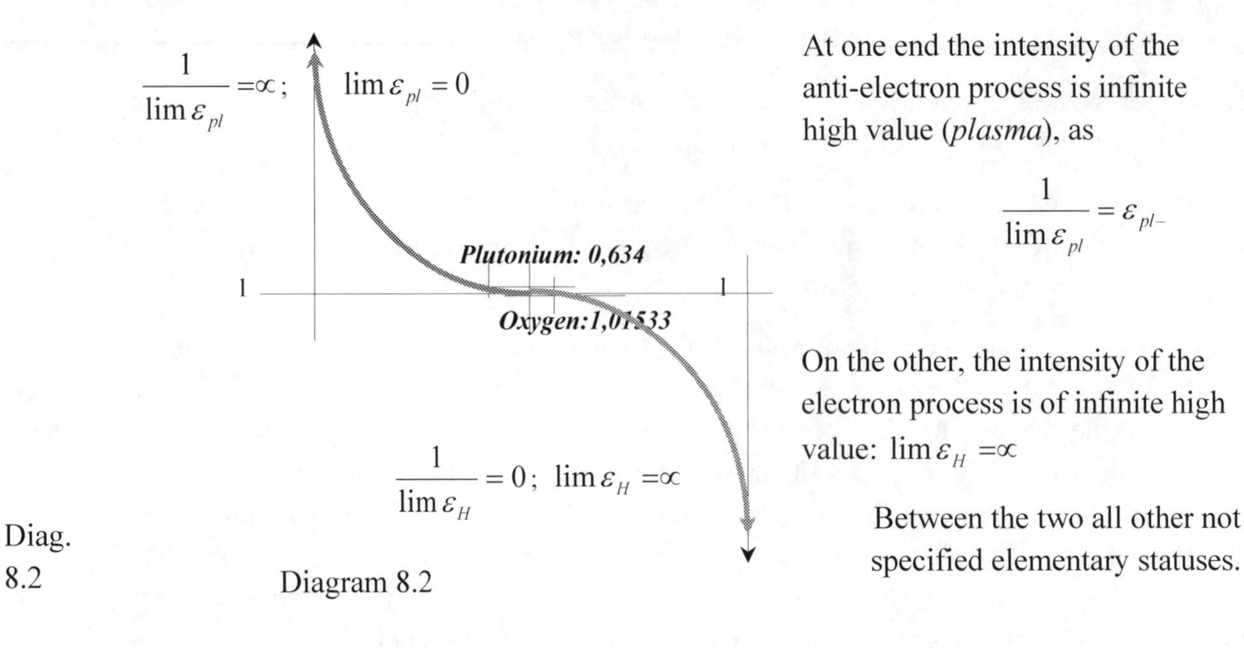

Diag. 8.2

Diagram 8.2

At one end the intensity of the anti-electron process is infinite high value (*plasma*), as

$$\frac{1}{\lim \varepsilon_{pl}} = \varepsilon_{pl-}$$

On the other, the intensity of the electron process is of infinite high value: $\lim \varepsilon_{H} = \infty$

Between the two all other not specified elementary statuses.

The plasma in the centre and the solid elementary statuses on the hardened surface, result of the cooling-expansion, exist in parallel. Parallel intensity statuses mean different time counts in parallel; parallel existence of different space-times and different aggregate statuses within the space-time matrix.

The hardened solid surface of the plasma, the surface of the *Earth* has its particular quantum speed value and intensity. The change, the elementary evolution, however does not stop at this stage. The decrease of the quantum speed and of the intensity continues and the elementary evolution generates elementary processes not just in liquid and gaseous aggregate state, representing the externally generated conflicting statuses under the quantum impact of gravitation, but also without the formulation of complete elementary cycles.

The difference in the aggregate status of the elementary processes depends on the intensity of the conflict between the elementary processes. Each elementary process has its space-time. The time count is function of the quantum speed and the intensity values.

The diapason from the quantum speed value on the *Earth* surface of 299,792 km/sec to the infinite low quantum speed of the *Hydrogen* process is infinite large. The field on the *Earth* surface is communicating, composing vegetation and living systems. The quantum treasure of the *Earth* surface/core uses its quantum impact in order to produce balanced elementary structures.

The matter is about events, matrix of quantum impact, elementary processes, and signals – *matrix of quantum information.*

The two ends of the *matter* are: the *Hydrogen* process with quantum impacts of infinite low intensity and the *plasma* with quantum impacts of infinite high intensity.

8.2 *Gravitation,* the formulation of the *matter*

S. 8.2

Elementary processes represent infinite large number of elementary cycles at different stages of the progress – at certain quantum speed and intensity values. The diapason of all known elementary processes extends from electron process intensities of $\varepsilon_{tr-Pu} = 0{,}6$, the *trans-Plutonium* processes to $\varepsilon_{He} = 1{,}016$ the *Helium* process, with quantum speed values, the range from 299792 km/sec to 365000 km/sec. The intensity of the *plasma* is $\lim \varepsilon_{pl} = 0$; the intensity of the *Hydrogen* process is $\lim \varepsilon_x = \propto$.

Solid subjects mean infinite number of operating in parallel elementary cycles at different stages of the elementary progress. The original anti-electron process *blue shift* quantum conflict of the plasma with infinite short time count is cooled down to the aggregates status called "solid" with increased time count of the process and without electron process surplus. Liquid and gaseous statuses represent conflicting stages with the surplus of the electron processes.

The *cooling* itself is no other than the quantum impact of gravitation. The higher the surplus of the generating anti-electron process quantum drive is, the higher is the quantum impact of gravitation

The cooling relates to the conflicting anti-electron processes, with certain IQ_{x-} anti-drive value of the elementary process.

$$e_{x\Sigma-} = n\frac{dmc_x^2}{dt_i\varepsilon_{x-}}\left(1-\sqrt{1-\frac{(c_x-i_x)^2}{c_x^2}}\right); \qquad \text{8B1}$$

The cooled down IQ anti-drive establishes the intensity of the inflexion and the intensity of the proton process:

$$IQ_{x-} = n\frac{c_x^2}{\varepsilon_{x-}} - e_{gr} = \frac{c_x^2}{\varepsilon_{x-}}; \qquad \text{8B2}$$

The aggregate value of the IQ anti-drive in 8B2 will be of less value, corresponding to the $IQ_{x-} = const$ elementary rule. With the decrease of the surplus of the anti-electron processes. The quantum speed will be of less and less value, the space-time of the elementary process becomes smaller and smaller.

$$IQ_{x-o} = n_\Sigma\frac{c_x^2}{\varepsilon_{x-}} - n_{gro}\frac{c_x^2}{\varepsilon_{x-}} = n_{e1}\frac{c_{x1}^2}{\varepsilon_{x-1}}; \qquad IQ_{x-1} = n_e\frac{c_{x1}^2}{\varepsilon_{x-1}} - n_{gr1}\frac{c_{x1}^2}{\varepsilon_{x-1}} = n_{e2}\frac{c_{x2}^2}{\varepsilon_{x-2}}; \qquad \text{8B3, 8B4}$$

Developing 8B3 and 8B4 to *n* elementary cycles:

$$IQ_{x-n} = n_\Sigma\frac{c_x^2}{\varepsilon_{x-}} - n_{gro}\frac{c_x^2}{\varepsilon_{x-}} - n_{gr1}\frac{c_{x1}^2}{\varepsilon_{x1}} - n_{gr2}\frac{c_{x2}^2}{\varepsilon_{x2}} - \ldots - n_{gr(n-1)}\frac{c_{x(n-1)}^2}{\varepsilon_{x(n-1)}} = n_{en}\frac{c_{xn}^2}{\varepsilon_{x-n}} \qquad \text{8B5}$$

Calculating only with the numbers of the anti-electron process impacts of gravitation the equation below should be valid:

$n_\Sigma - n_{gro} - n_{gr1} - n_{gr2} - \ldots - n_{gr(n-1)} - n_{grn} = 0$; and

$n_\Sigma = n_{gro} + n_{gr1} + n_{gr2} + \ldots + n_{gr(n-1)} + n_{grn}$

In this case $\frac{c_x^2}{\varepsilon_{x-}} = const$ indeed 8B6

as it always was supposed!! 8B7

Ref. 8B6

The equations on the previous page also prove the decreasing character of the number of the parallel elementary cycles of the evolution. The closer the elementary process to the plasma status is, the higher is the number of its parallel elementary cycles.

Gravitation is work impact. This is the quantum impact of the plasma and the elementary evolution through the space-times of the progress. Work impact obviously is the anti-electron process quantum drive of the anti-proton process as well. This is the reason the process with time count $\lim t_i = \propto$ has its end and the new cycle can start.

The temperature is the adequate characteristic of the conflict, generated originally by the anti-electron processes of the plasma and by the close to plasma elementary processes of the evolution. The cooling (work), consequence of the anti-electron process conflict results in quantum impact and generation of liquid and gaseous statuses on the *Earth* surface and within regions close to the surface *Earth.*
Warming/heating up elementary processes the conflict is increasing and the aggregate status may be changed. From solid to liquid or gaseous.
The intensity of the anti-electron and the electron processes close to the *plasma* status is:

$\varepsilon_{x-} >> 1$ and this way $\varepsilon_x << 1$.

8C1

As $\frac{c_x^2}{\varepsilon_{x-}} = const$; the number of the formulating *inflexions*,
with reference to 3F3 is: $n_{x+1} = n_x\left(1 - \frac{\varepsilon_{x-} - 1}{\varepsilon_{x-}}\right)$;

The conflict at the anti-side is infinite high; the relative number of the elementary cycles going through the *inflexion* is increasing with the decrease of the number. The less the value of n_x is, the higher is the proportion of the *inflexion.*

8C2

Closer and closer to the *Hydrogen* process, the number of the elementary cycles, going through the *inflexion* is increasing,
as with reference to 3F6: $n_{y+1} = n_y\left(1 - \frac{\varepsilon_y - 1}{\varepsilon_y}\right)$;

The higher the value of n_y is, the more is the number of the inflexions.

The solid surface of the *Earth* means neutron process potential, with coefficients of electron process intensities of $\varepsilon_y < 1$ and $\varepsilon_y << 1$, without completed elementary processes. Starting from the *plasma* status of infinite high anti-electron process surplus and conflict, the *Earth* surface represents elementary processes with close to equilibrium status and without completed elementary cycles. The completion of the neutron process results in anti-neutron process potential and the appearance and the growth of vegetation.

Everything starts with the quantum impact of the plasma.
The quantum impact of the plasma, as quantum signal or information has different forms of its appearance. The intensity of the information decides the format of the appearance. Statuses, close to plasma are in high temperature conflicting stages. The quantum impact of the processes of the cooled down statuses represent solid or liquid or gaseous aggregate stages, but still of the same information, born in the plasma conflict.

Elementary processes mean energy transfer and quantum impact. Anti-electron processes mean the quantum impacts of gravitation. Electron processes as drives mean the quantum impact of the elementary communication. The difference in the communication is within the space-time of the process. In other words the time count and the quantum speed of the processes. The information is establishing the *matter* in space and time.

If we take that the plasma is conflict of quantum impacts, signals of infinite high intensity, result of the collapse of quantum impacts, signals of infinite low intensity, our all known elementary processes are quantum impacts, signals in conflicting stages between these two ends. Quantum impacts and signals happen in time and represent frequencies. Quantum signal in time are information. Information on the number of the impacted quantum and on the quantum speed and the intensity of the impact.

All around us in our life is *blue shift* quantum impact of electron processes of elementary processes, naturally generated like the *Sunshine* or the background radiation of gravitation, or generated artificially, like all radio signals, propagating in the quantum space of quantum impulses.
Electron process is quantum impact of certain value, the intensity of extension, the driving impact of conflicts and the collapse.
The *quantum* impact of the anti-electron process is escorted by the intensity of gravitation, the *mechanical* impact of the extension.

The extension in its global terms, *gravitation* is the consequence of the parallel existence of the separate space-times of elementary processes.
Plasma is of infinite high intensity, infinite high quantum speed – a process in a space-time of unlimited, infinite large value. And this unlimited space-time has been surrounded by infinite large number of elementary processes of less and less intensity and quantum speed – result of the cooling process of the plasma. There is the plasma drive inside, with the space-time of the plasma, containing the space-time of all elementary processes.

The one and the only mechanical impact of the elementary process is the *inflexion*, where the collapse turns into expansion. This is an impact in our conventional terms, which has significant physical impact. The inflexion is the point where the whole energy/capacity/intensity of the process is generating. The collapse is the increase of the intensity, driven by external source. The expansion is the process on its own. The whole energy is formulating and generating in *inflexions*!

The quantum speed is the intensity of the change itself. All speed values mean change in time, just the subjects of the change are different. The quantum speed is the measurement of the intensity of the change within the *inflexion*. After the inflexion the intensity of the impact is decreasing. The gradient of the decrease varies, establishing by this the difference between elementary processes. The remaining impact of the electron process is representing the inflexion through the quantum speed value. And the change is reformulating from expansion to collapse and inflexions happen again and again.

S. 8.3

8.3
Soil and ash

The plasma state means the full conflict of the anti-electron processes.
The cooling down of the plasma results in the hardened core of the *Earth*, full with minerals of elementary processes. The soil, the status of the *Earth* surface is specific: this is the internal burnout of the plasma. While it is with massive loss on the intensities of the electron processes, soil is full of elementary potential, jut with not completed elementary cycles.

Picture 8.1 shows the remains of the crumbling lava at the Hawaii island. The internal burnout of the plasma means, the conflict is generated and fed by the intensity of the plasma itself. This way the burnout has its limits. After a point the conflict stops, as the intensity of the conflict has been lost. The remains are with electron process impacts and potential elementary communication.

Pic. 8.1

Picture 8.1

The soil on the surface means, the quantum speed is of less value and the intensity of the lava burnout (result of the loss of the intensity of the conflict) is still higher than the quantum speed and the intensity at the *Earth* surface.

The quantum speed of the soil is less than the quantum speed of the *Helium* process, the last completed elementary process of the Periodic Table. The value of the intensity coefficient of the electron processes of the soil obviously less than 1, as there is no intensity surplus generation within the soil: the anti-proton/proton inflexion is missing.

The soil is full of energy and intensity capacities, but without identified elementary systems. It is ready for elementary communication with vegetation. This communication can be initiated and enforced by the electron process surplus of the water. The communication with vegetation starts even without any human contribution. Picture 8.2 demonstrates the results of the communication on the stone hard solid lava remains on the Big Island of Hawaii with other still operating volcanos. The vegetation is growing on the remains of the volcano, on the lava remains of 40 years old.

Pic. 8.2

Picture 8.2

The not completed elementary processes of the soil is welcoming external communication, which helps to complete the cycle. Water is with electron process surplus.

As result of the quantum communication of the soil with water and the surplus of the electron process *blue shift* impact, available within the water, there is a generation of quantum membrane "pressure", necessary to increase the intensity of the anti-electron process of the soil to complete the elementary cycle. The elementary cycle is completing, with the benefit of the generation of the energy potential of the proton process for use – actually by vegetation!
Vegetation has its own intensity potential to be utilised in quantum communication.

If, minerals or soil become part of external electron process *blue shift* conflict, fire, the impact of the conflict – as driven externally – will not stop reaching the limits of the capacities of the soil, result of the elementary evolution. It will be fully burning out all existing in the soil electron process intensities. Ash remains without communication capacities.

The radically reduced value of the *IQ* drive within the ash makes it similarly not completed, and because of the practically not driven neutron process – its weight is light.

$$IQ_{ash} = \frac{c_{ash}^2}{\varepsilon_{ash}}; \quad c_{ash} << c_E; \quad \varepsilon_{hard} < 1; \quad \text{and} \qquad \text{8D1}$$

The ash is of light weight, in easy conflict with gravitation, quantum speed of limited value, with slight remaining potential to neutron process dominance. *Ash* means burnout, loss of the intensities, results of a conflict that happens on the *Earth* surface.

8.4 *Matter* is the matrix of information

S. 8.4

Conflicting quantum signals (impacts) of specific frequency (intensity) create high frequency spots, area with increased temperature. The intensity of the generating conflict depends on the number and the frequency of the conflicting signals.

The intensity concentration of the information in Diagram 8.3/a on the next page is of increased value.
The same number of the conflicting information in 8.3/b is losing on its intensity and becomes of different status. The incoming information is the same, but the resulting information in the centre is less: the frequency is of less value.

Matter is a process with infinite variety of intensities.
Matter means the conflict, which establishes its appearance, density and aggregate status: solid, liquid, gaseous or even plasma.
Matter is not about particles rather processes. *Matter* exists in time. There is no time count without the event of the change of the *matter*.

The quantum signals in the centre create high frequency quantum impact function of the number of the quantum impacts

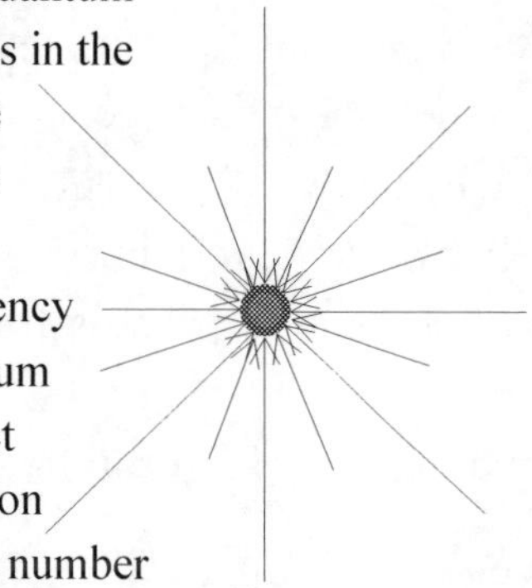

cooling

Keeping the conflict on, but cooling it,

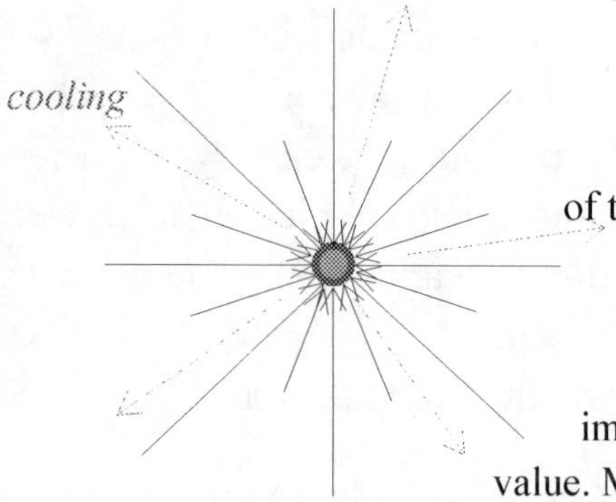

the *aggregate* status of the conflict becomes different: The intensities of the repelling impacts will be of less value. Might the conflict be disappear. As the volume of the incoming and the leaving intensities are similar

Diag. 8.3

Diagram 8.3/*a,b*

Matter is quantum impact and quantum signal – in simple terms: information!

Ch.9

9
The free energy is with us

Our life is about permanent "fight" for new energy sources. The cooling process of the *plasma* is always with us: As *gravitation* proves it, the energy source is given. The acceleration of the *Hydrogen* process and its conflict with the quantum impact of *gravitation* gives the chance to use it.

Ref. QISM S.6

Chapter 6 of the book of *The Quantum Impulse and the Space-Time Matrix* gives the basics of the energy generation. Explanation in details, measured data and the description of the small scale experiment follow in this section.

S. 9.1

9.1
Introduction, short explanatory note about the principles

The purpose of the acceleration of the *Hydrogen* process is the increase of the intensity of the process. Acceleration creates an additional conflict with the quantum impact of *gravitation.* Additional, because the *Hydrogen* process has already been intensified on the *Earth* surface by the quantum impact of *gravitation*, increasing this way the intensity of its gaseous state: The quantum speed of the *Hydrogen* process corresponds to the speed of light on the *Earth* surface.

The *blue shift* quantum impact of the electron process of the normal *Hydrogen* process on the *Earth* surface is:

$$e_H = \frac{dmc_E^2}{dt_E \varepsilon_H}\left(1-\sqrt{1-\frac{(c_E - i_E)^2}{c_E^2}}\right); \qquad (1) \qquad \text{where:}$$

9A1

c_E – the speed of the quantum communication (the speed of light) on the *Earth* surface;

$\lim i_E = c_E$ – the speed of the operation of the electron process of the *Hydrogen* process of increased intensity;

$\lim \varepsilon_H = \infty$ – the intensity coefficient of the electron process of the *Hydrogen* process,

which means its intensity is of infinite low value, as $\varepsilon_H = \frac{\varepsilon_p}{\varepsilon_n}\sqrt{1-\frac{(c_H - i_H)^2}{c_H^2}}$,

9A2

where $\lim \varepsilon_n = 0$, the intensity of the neutron process, infinite low value – this is the reason why the neutron process of the *Hydrogen* process cannot be measured;

ε_p marks the intensity of the proton process.

dt_E – is the time count on the *Earth* surface, corresponding to $\lim i_E = c_E$, the speed of the sphere symmetrical expanding acceleration of the *Earth* surface.

The acceleration of the *Hydrogen* process to speed v means the intensity of the electron process will be further increased, to value of:

9A3
$$e_{Hv} = \frac{dmc_E^2}{dt_E \varepsilon_H \sqrt{1-\frac{v^2}{c_E^2}}} \left(1-\sqrt{1-\frac{(c_F-i_F)^2}{c_F^2}}\right);$$

The quantum impact of *gravitation* is the quantum impact of the anti-electron processes of the chain of the "from *plasma-to the Hydrogen* process" elementary evolution:

- All elementary processes, starting from the *plasma* and ending with the *Hydrogen* process have their own specific to the process quantum speed and electron process intensity values. The quantum speed and the intensity of the electron process of the *plasma* are of infinite high values: $\lim c_{pl} = \propto$, $\lim \varepsilon_{pl} = 0$. The *Hydrogen* process is the other end: the parameters of the *Hydrogen* process are of infinite low values: $\lim c_H = 0$ and $\lim \varepsilon_H = \propto$.
- The quantum speed of the *Hydrogen* process on the *Earth* surface corresponds to the quantum speed value on the *Earth* surface; as it has already been accelerated by the conflict with *gravitation* – the sphere symmetrical expanding acceleration of the *Earth* at $\lim i_E = \lim g\Delta t = c_E$ constant speed. The same way as the quantum speed values of the *Helium*, the *Nitrogen* and the *Oxygen* processes have also been increased.
- The infinite wide range of the elementary evolution is well presented by the period of our known elementary processes between the *Plutonium* process with electron process intensity coefficient of $\varepsilon_{Pu} = 0{,}648$ and the *Helium* process with electron process intensity coefficient of $\varepsilon_{He} = 1{,}01389$.
- The diapason of the intensity values from the *Plutonium* process to the *plasma* and from the *Helium* process to the *Hydrogen* process mean in fact infinite wide intensity variations, with the increase to infinite high values in the direction towards the *plasma* and with the decrease to infinite low values towards the *Hydrogen* process – giving the proof of the wealth of the energy content of the *Earth.*

The numbers of the parallel elementary cycles of the elementary processes are different. This is the reason that the anti-direction for the majority of the elementary processes is with anti-electron process surplus. The *H, He, Ni, O, Si, S* and the *Ca* processes are the exemptions, where the surplus is accumulating on the direct process side.

The *blue shift* impact of the anti-electron processes of all elementary processes of the *Earth,* starting from the plasma is the quantum impact of *gravitation,* measured on the *Earth* surface, as background radiation, value of 0,13-0,18 μ *Sv/h.*

The quantum impact of *gravitation* on the *Earth* surface is continuous and quasi equal. Where the continuity has not been established yet, the surface has been covered by water, which slows down the quantum speed of the quantum impact and makes the quantum communication quasi uniform above the *Earth* surface.

The quantum impact of gravitation is: $e_g = \frac{dmc_E^2}{dt_i \varepsilon_{E-}}\left(1-\sqrt{1-\frac{(c_E - i_E)^2}{c_E^2}}\right)$; 9A4

ε_{E-} is the intensity coefficient of the anti-electron process; its value is reciprocal to the intensity coefficient of the electron process. The value of the intensity coefficient for the *Earth* surface is equal to $\varepsilon_E = 1$ and this way the intensity coefficient of the anti-electron process is also equal to $\varepsilon_{E-} = 1$. This value has been taken as freely selected and corresponds to the intensity coefficient of the elementary process of the *diamond,* the *Carbon* mineral; demonstrating a kind of balanced status.

The quantum impact of *gravitation* is equal for all elementary processes as the quotient of the quadrat of the quantum speed and the intensity coefficient of the anti-electron process for all elementary processes are equal: $\frac{c_x^2}{\varepsilon_{x-}} = const$ 9A5

Ref. Table 3.1

In other words, the quantum speed value of the plasma is slowing down, the intensity values are decreasing and the number of the elementary cycles, with reference to Table 3.1, become less and less during the elementary evolution.

The time counts for all electron processes are identical, as the $\lim i_x = c_x$ operation gives equal results for all quantum speed values: $dt_i = \frac{dt_o}{\sqrt{1-\frac{i_x^2}{c_x^2}}}$; 9B1

dt_o in this expression gives the status of rest indeed.

The identical $\lim t_i = \infty$ time count of the electron processes is the precondition of the elementary communication of the elementary world.

Ref 9A1 9A2 9A3

With reference to 9A1 and 9A2, the already existing conflict establishes the "normal gaseous" state of the *Hydrogen* process on the *Earth* surface.
With reference to 9A3, the acceleration creates additional conflict, influencing the mechanical impact of gravitation – the weight impact of the *Hydrogen* process.

Gravitation is the sphere symmetrical expanding acceleration of the *Earth* at constant $\lim i_E = \lim g\Delta t = c_E$ speed and constant *g* acceleration. There is no reason questioning this statement and this statement does not contradict to any practical findings.

The definition of *gravitation* as sphere symmetrical expanding acceleration of the *Earth* at constant $\lim i_E = \lim g\Delta t = c_E$ speed gives simple and consistent explanation to all our findings. The sphere symmetrical expanding acceleration is not just one of our technical reciprocities, but together with the sphere symmetrical expanding collapse they give one of the most important elementary pairs: *pulsation.*

The constant acceleration of the *Hydrogen* process results in constant conflict with the quantum impact of *gravitation* within the channel of the acceleration.
The sphere symmetrical expanding acceleration of the *Earth*, the *mechanical impact of gravitation* is the one resulting in the weight of the *Hydrogen* process within the channel.

The developing conflict between the *quantum impact of gravitation* and the speeded up *Hydrogen* process modifies this mechanical impact, the weight of the *Hydrogen* process within the channel.

With the freedom of the channel to move up and down, the acceleration results in the modification of the weight of the channel with the *Hydrogen* process. The higher the conflict with the quantum impact of gravitation is, the less is the weight of the construction. The increased intensity of the electron process of the *Hydrogen* process withstands the quantum impact of gravitation.
Weight increase in the contrary, demonstrates the weakening of the conflict,

With the acceleration impact permanent, and the quantum impact of gravitation permanent, the developing conflict is also permanent:

9B2 $$\frac{dq_g}{dt_E} \pm \frac{dq_H}{dt_E} = \mp \Delta G;$$

Where

q_g means the quantum impact of gravitation (*qig*) and q_H the quantum impact of the *Hydrogen* process in acceleration, ΔG is the modification of the actual weight value.

If the construction of the channel of the acceleration has been hanged up, or has the chance in any other way to influence the appearance of the weight, the conflict finds its balancing status. The work intensity benefit of the acceleration of the *Hydrogen* process results in a "lifting up" or a "letting down" change. Until a certain generated conflict is acting and finds its balance, the relation with the gravitation is also constant – the weight is constant.

If the channel of the acceleration is fixed – built under or above the *Earth* surface, but certainly fixed, with no chance to compensate the conflict by the motion of the construction – the balancing release is also missing. In this case the conflict is escalating. The externally generated conflict by the acceleration becomes internal, "closed" within the channel.

There is no way in this case to release the conflicting impact within the channel. The acceleration is permanent, the channel has been fixed –the conflict has been closed. The deepening conflict of the *Hydrogen* process with the quantum impact of gravitation results in electron process conflict = heat generation.
And the heat generation in our conventional understanding is energy to use.

A *Hydrogen* process accelerator of industrial scale means an accelerating channel built without free motion and with the acceleration of the *Hydrogen* process close to the speed of light. There is an accumulation of conflict between the quantum impact of gravitation and the quantum impact of the *Hydrogen* process within the channel of the acceleration.
This conflict generates heat!

There is no way in normal circumstances to prove the heat generation benefit of the acceleration of the *Hydrogen* process. The only chance is the demonstration of the weight impact of the acceleration.

The *Hydrogen* process is the only one which can be used for this purpose: the acceleration is increasing the intensity of the process, but the process remains of infinite length as it was in its normal natural status without the acceleration. (All other elementary processes would result in the completion of the elementary process with increased intensity.)

Ref. QISM S.6.4

With reference to Section 6.4 of *"The Quantum Impulse and the Space-Time Matrix"* the industrial application needs the accelerating channel, built

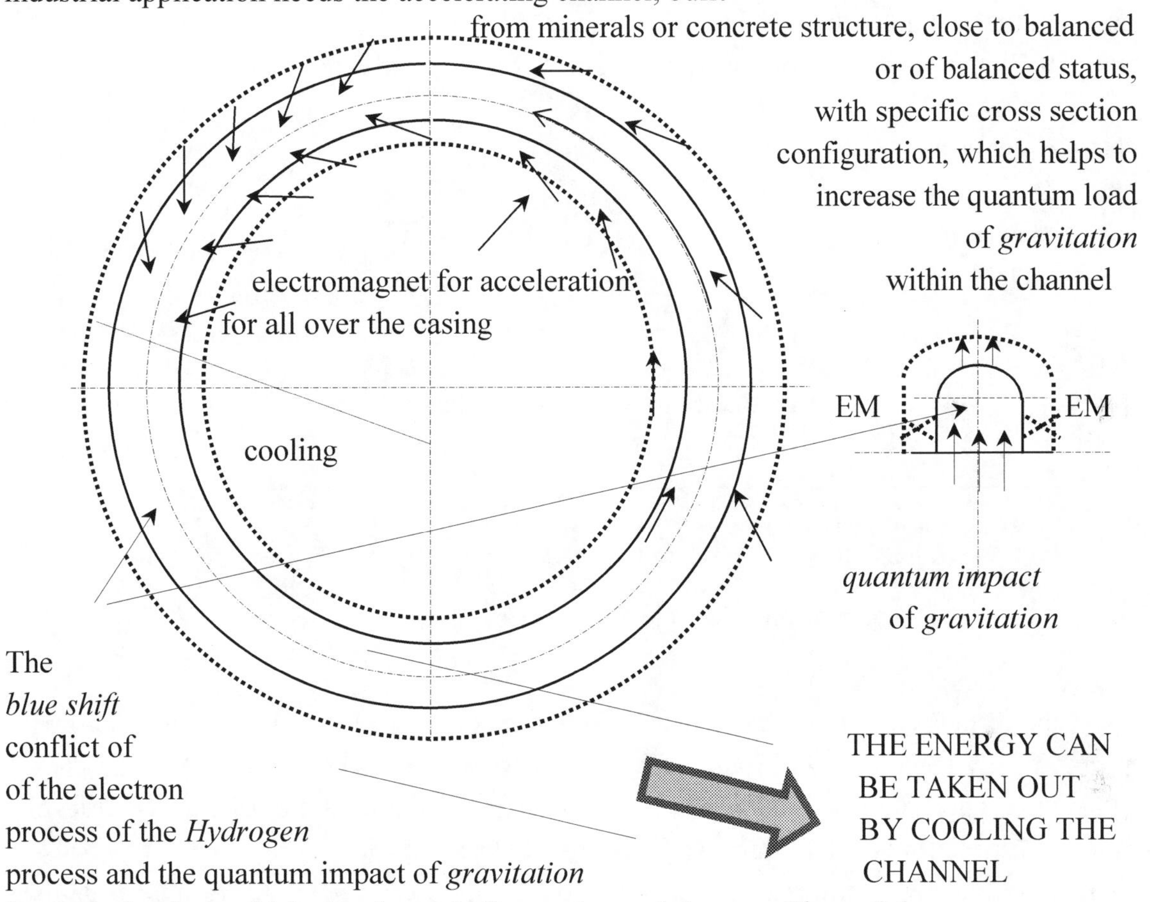

Figure 9.1

Fig. 9.1

S_c e_s S_b e_b S_E

The quantum impact of *gravitation* and the *Hydrogen* process

with increased intensity inside are conflicting

Figure 9.2

$S_b = S_E$; and $S_c > S_b$

therefore the intensity of the flow of the quantum impact of *gravitation* through the basic surface shall be higher, meaning the intensity coefficient shall be of less value: $\varepsilon_b < \varepsilon_c$ – as the continuity of the quantum impact through the cover surface to the environment shall be provided.

In this case: $e_{basis} = \dfrac{e}{S_b \varepsilon_b}$; and $e_{cover} = \dfrac{e}{S_c \varepsilon_c}$;

$e_b = e_c = e$ the quantum energy transfer in absolute volume through the basic and the cover surfaces are one and the same. But the intensity of the energy transfer through the basic surface is higher: $e_{cov} < e_{basis}$

Fig. 9.2

S. 9.2

9.2 Examples of the small scale *Hydrogen* process acceleration experiment

The objective of the experiment is to give proof and demonstrate that the conflict of the quantum impact of the *Hydrogen* process in acceleration and the quantum impact of gravitation has its influencing effect on the mechanical impact of *gravitation*: The conflict results in the change of the weight.

Ref. 9A3

Figure 9.3 below shows the principal components of the small scale experiment:

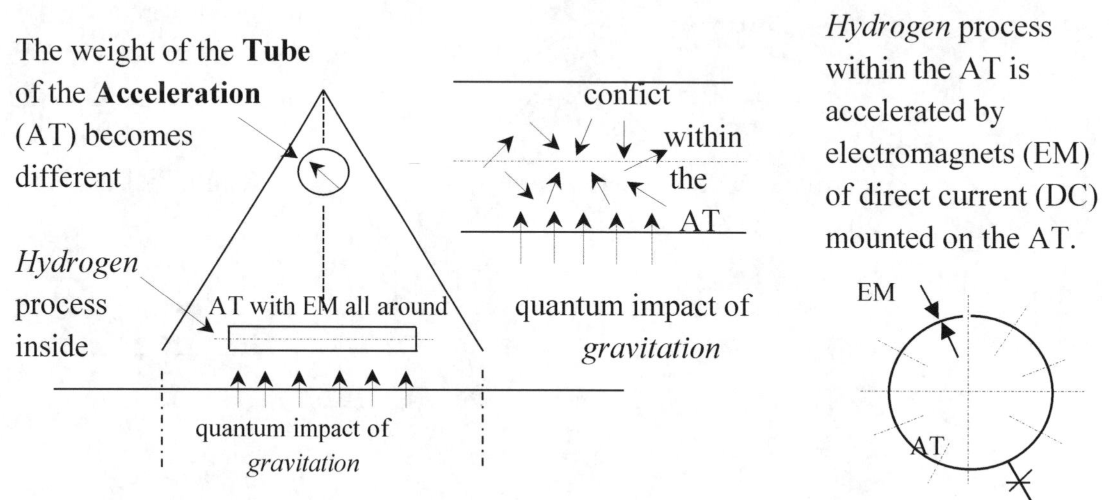

Fig. 9.3

AT- Tube of the Acceleration, EM - electromagnet

Figure 9.3

The Tube of the Acceleration is of *ND 20* mm at a diameter of the circle of *900 mm*, with approximate volume of *0.9 l*. The *Hydrogen* process in the channel is *99,9 %* clean and of *4,9 bar* pressure. There are electromagnetic coils (EM) all around the AT in several levels, winded in special way. The electromagnets drive the acceleration.
There were hundreds of measurements made during the experiment. All results strengthen the finding: the quantum impact of the *Hydrogen* process in acceleration is in conflict with the quantum impact of *gravitation*. The conflict is influencing the mechanical impact (the measured weight) of the accelerating system.
The most representative 4 examples are presented in this section.

Measuring the weight impact of *0.9 l Hydrogen* process in acceleration is not an easy task. The measurements first were made by a simple balance, but measuring a change less than *1g*, is difficult. The results were supporting the expectations on the loss on the weight of the *Hydrogen* process, but adequate proofs needed more precise approach.

This was the reason of changing the methodology and instead of measuring the direct weight, the effect of change of the weight was measured. A two-armed balance was constructed and the movement of the longer arm was the subject of the measurement. The positions of the arm during the acceleration were recorded and the videos and the pictures taken later assessed.

9.2.1 Additional principal notes to the experiment

S. 9.2.1

The acceleration increases the *IQ* drive of the *Hydrogen* processes.
The heating up also results in the intensity increase of the *IQ* drive of elementary processes. If there is no room for working out the intensity increase, the heating up results in internal conflict. The acceleration of the *Hydrogen* process, as result of the impact of the electromagnets, also causes temperature increase.
If we want to specify the intensity growth of the *blue shift* quantum impact of the electron processes and separate the two, we can write

$$\frac{dmc_x^2}{dt_i\varepsilon_x\sqrt{1-\frac{v^2}{c_x^2}}}\left(1-\sqrt{1-\frac{(c_x-i_x)^2}{c_x^2}}\right)=\frac{dmc_x^2}{dt_i\varepsilon_x\sqrt{1-\frac{v_a^2}{c_x^2}}\sqrt{1-\frac{v_h^2}{c_x^2}}}\left(1-\sqrt{1-\frac{(c_x-i_x)^2}{c_x^2}}\right); \quad \text{9C1}$$

where v_a corresponds to the impact of the acceleration and
v_h to the heat impact of the coils of the accelerating electromagnets.

The difference between the two impacts is, that
during ***heating up,***

- the source of the developing internal *blue shift* conflict is *external* (heat);
- the conflict results in the increase of the intensity of the neutron collapse,

$$\frac{dmc_x^2}{dt_i\varepsilon_x\sqrt{1-\frac{v_h^2}{c_x^2}}}\left(1-\sqrt{1-\frac{(c_x-i_x)^2}{c_x^2}}\right)=n_h\frac{dmc_x^2}{dt_i\varepsilon_x}\left(1-\sqrt{1-\frac{(c_x-i_x)^2}{c_x^2}}\right) \quad \text{9C2}$$

which is equivalent to the increase of the number of the electron process quantum impacts.

It can also be written as $n_h\frac{dmc_x^2}{dt_i\varepsilon_x}=IQ_{xh}$; an increased quantum drive, corresponding to the conflict. 9C3

- the conflict results in increased intensity of the neutron/anti-neutron inflexions and increases the number of the generating anti-electron processes,
  - in the case of $\varepsilon_x > 1$ (*H, He, C, Ni, O, Si, S, Ca* processes) this increases the number of the anti-proton/proton inflexions for the unit period of time, with the increase of the number of the proton processes, but of the intensity of their own elementary processes. The source of the intensity increase is _external_, but the response from the anti-process is _natural_, with the increase in numbers, with the guarantee of the original elementary process.
  - with the heating on, even the elementary processes with $\varepsilon_x < 1$ (all other elementary processes) become liquid and even gaseous;
    The conflict on the direct side is of high intensity and the number of the neutron/anti-neutron inflexions results in increased anti-electron process generation with less proportions going for the quantum impact of gravitation. The proton process here also insures the natural characteristics of the elementary process.

- the heating up in limited space in conventional terms results in the increase of the elementary cycles, the temperature and the pressure.

  The heating up of an elementary process of constant volume (in a closed vessel) corresponds to a kind of mass change as well, just as the mass value is fixed the aggregate status of the elementary processes is the one, which is changing instead.
  As proof, all elementary processes have different thermal coefficient of their own specific value. The reason is that all elementary processes have their own quantum speed value, also of specific value.

There is no increase in the numbers of the elementary cycles in the case of the heating up of the *Hydrogen* process, since the elementary cycles have not been completed. The heating up results in conflict, but the consequences are different.

during ***acceleration***

- the impact is external as well, but the intensity increase of the process goes with the growth of the quantum speed: $c_{xa} > c_x$.

9C4
$$\frac{dmc_x^2}{dt_i\varepsilon_x\sqrt{1-\frac{v_a^2}{c_x^2}}}\left(1-\sqrt{1-\frac{(c_x-i_x)^2}{c_x^2}}\right)=\frac{dmc_{xa}^2}{dt_i\varepsilon_x}\left(1-\sqrt{1-\frac{(c_x-i_x)^2}{c_x^2}}\right);$$

- acceleration in conventional terms is a kind of *relative mass (weight) increase*, changing the space-time of the process and has its *relativistic impact*!

  The acceleration results in the increase of the *IQ* quantum drive, with the slowing down of the time count and with the increase of the speed of quantum communication: The space-time of the elementary process is widening.
  Without going deep into this subject, it shall be noted that while the relativistic impact, the increasing value of the quantum speed and the widening space-time of increased intensity are equally valid for the acceleration of the *Hydrogen* process as well, the *Hydrogen* process is an open process (as the plasma as well).
  While the increasing quantum drive in conventional terms results in mass (weight) increase, in the case of the *Hydrogen* process, the increasing quantum drive means increasing quantum impact. And the increasing quantum drive is the one in conflict with the quantum impact of gravitation. And it results in weight reduction.

Assessing the results, it is clear that there is no way the impact of the acceleration and the heating up could be separated. Once the acceleration (and with that the heating up) of the *Hydrogen* process is over, the increased thermal status will be lost by the natural cooldown of the system. The rehabilitation of the space-time of increased intensity and of the increased value of the speed of quantum communication happens in parallel with the slowdown of the flow.

The objective of the experiments was to demonstrate this weight decreasing impact.
The construction is simple, the level of the acceleration is far not significant, but the weight impact is measureable.

9.2.2 Results with the two-armed balance

S. 9.2.2

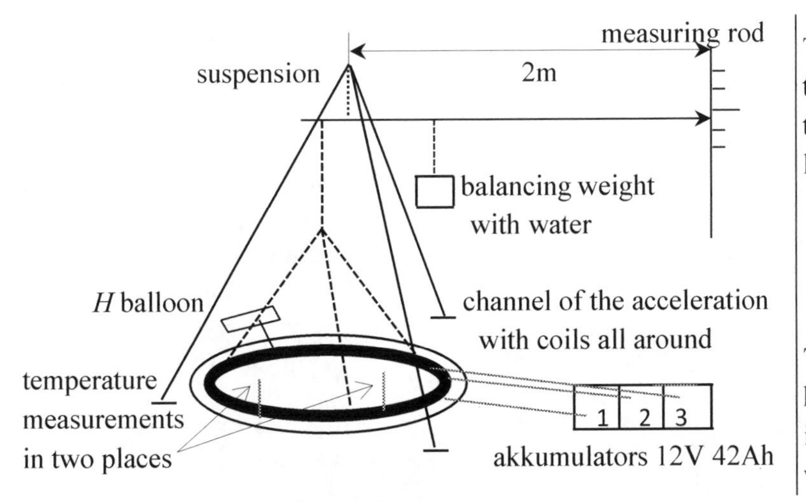

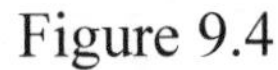

This is the figure of the two-armed balance, where the balancing weight of the longer arm is water.

Coils in 9 layers in one direction. Electricity is from 3 accumulators of 12 V of each.

The volume of the water has been selected for being in equilibrium with the weight of the system.

Figure 9.4

Fig. 9.4

Example 1.

Exp. 1

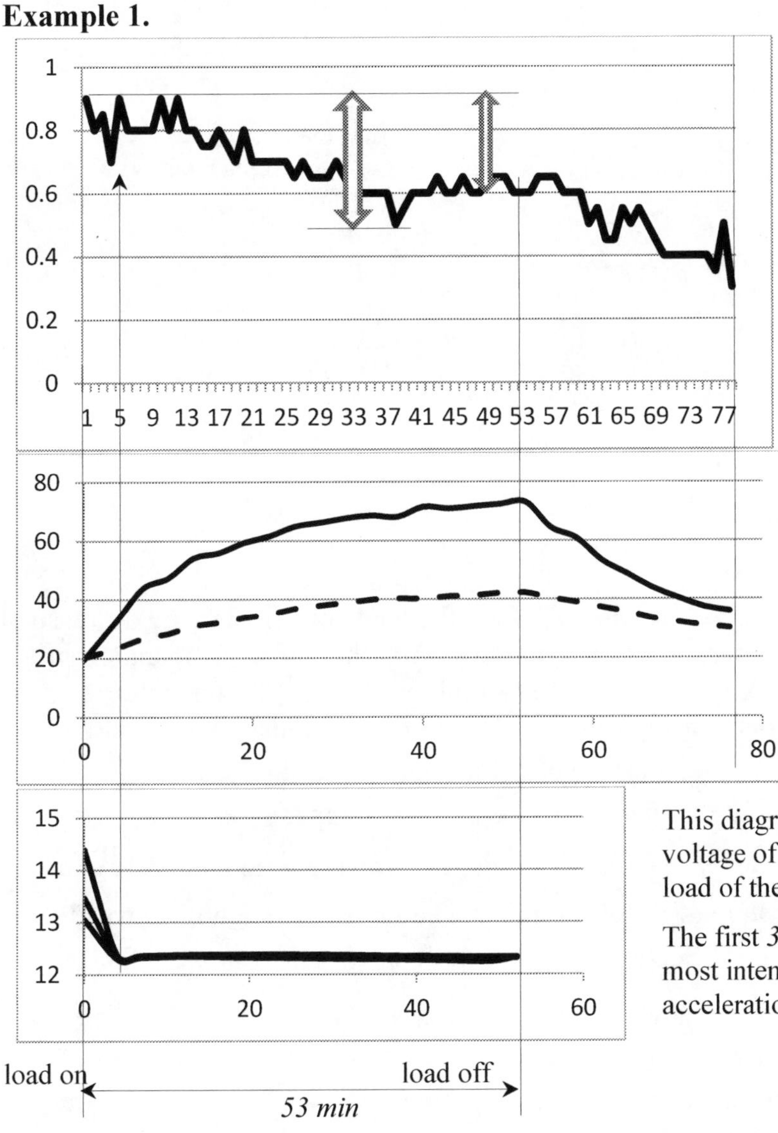

The maximal difference in vertical positions of the *2 m* long arm of the balance during the acceleration was: $\Delta = 0.9\text{-}0.5 = 0.4\ mm$

The weight reduction effect does not stop after the load of acceleration is off.

/The horizontal axis is the duration of the acceleration in *1 min* units./

This diagram is about the temperature of the electromagnets in ^{o}C in two specific spots.

The gradient of the increase of the temperature slows down after the first minutes of the load.

This diagram is about the change of the voltage of the *3* accumulators of the load of the electromagnets.

The first *3-5 minutes* of the load has the most intensive impact on the acceleration of the system.

Diagram 9.1

Diag. 9.1

The question was: Is this really the case that the reduction of the weight (the vertical motion of the longer arm of the balance) continues while the load has already been off? What is the reason of this?

The answer is coming from the results of the next examples.

Exp. 2

Example 2.

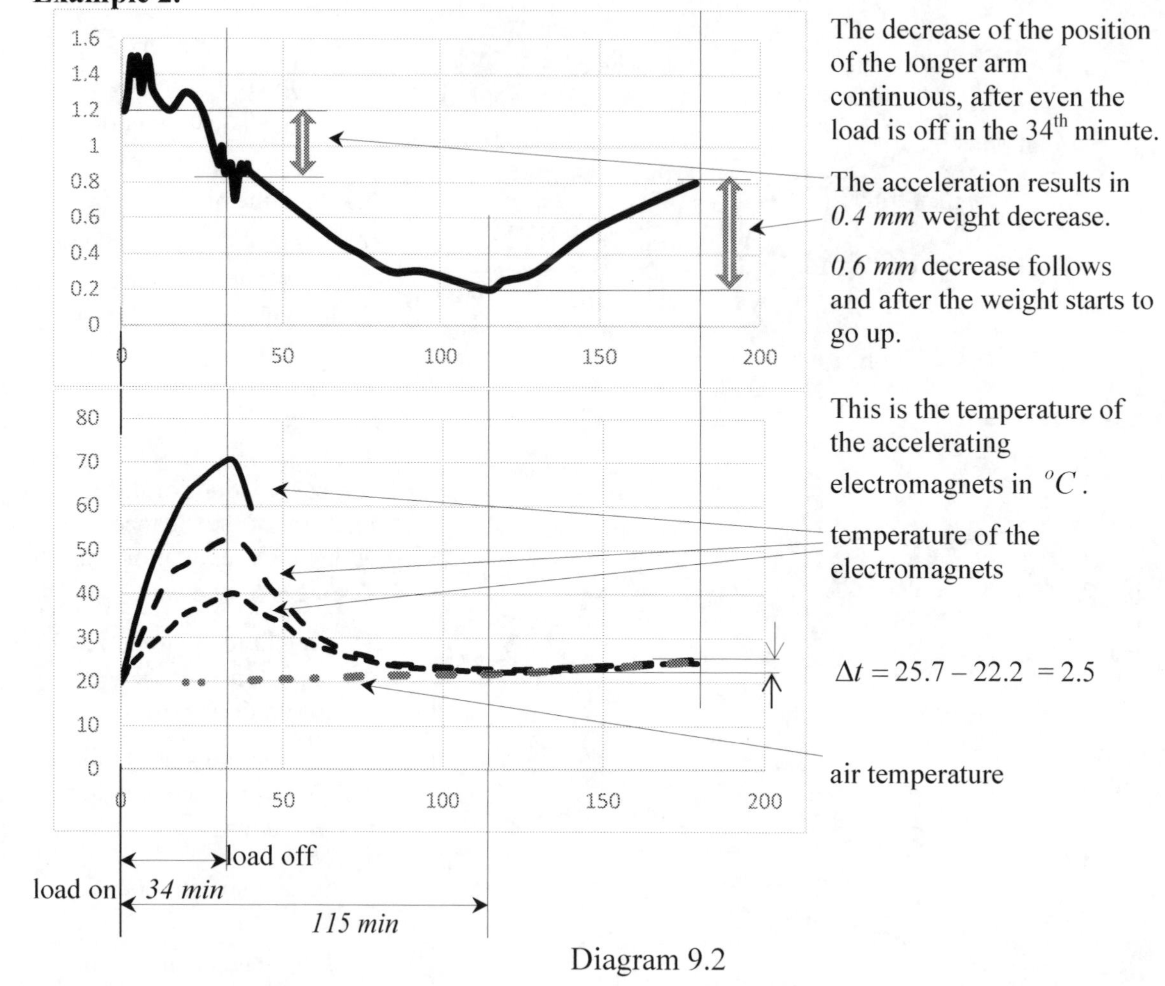

Diag. 9.2

Diagram 9.2

There was small disturbance at the 34th *minute*, when the load was off, but the decrease of the weight has continued. The reason of the continuation is the higher temperature of the accelerating system than the temperature of the environment. Once the temperature reaches the temperature of the environment (air) in the 115th minute, the decrease stops and the weight goes up. The earlier $\Delta = -0.6$ mm displacement for a decrease of more than $50\,^{o}C$ down goes back for $2.5\,^{o}C$ up for the increase of the external temperature.

The temperature of the *Hydrogen* process within the tube of the acceleration is equal or slightly more than the temperature of the electromagnets, in practice, slightly above $70\,^{o}C$. The temperature, with reference to Section 4.2, is electron process *blue shift* conflict.

Ref. S.4.2

There are still a couple of questions about the impact of the acceleration. The following examples try to give the explanations.

The most important question was: Is the (rehabilitating) weight increase reaching the original value, before the acceleration?

Example 3.

Exp. 3.

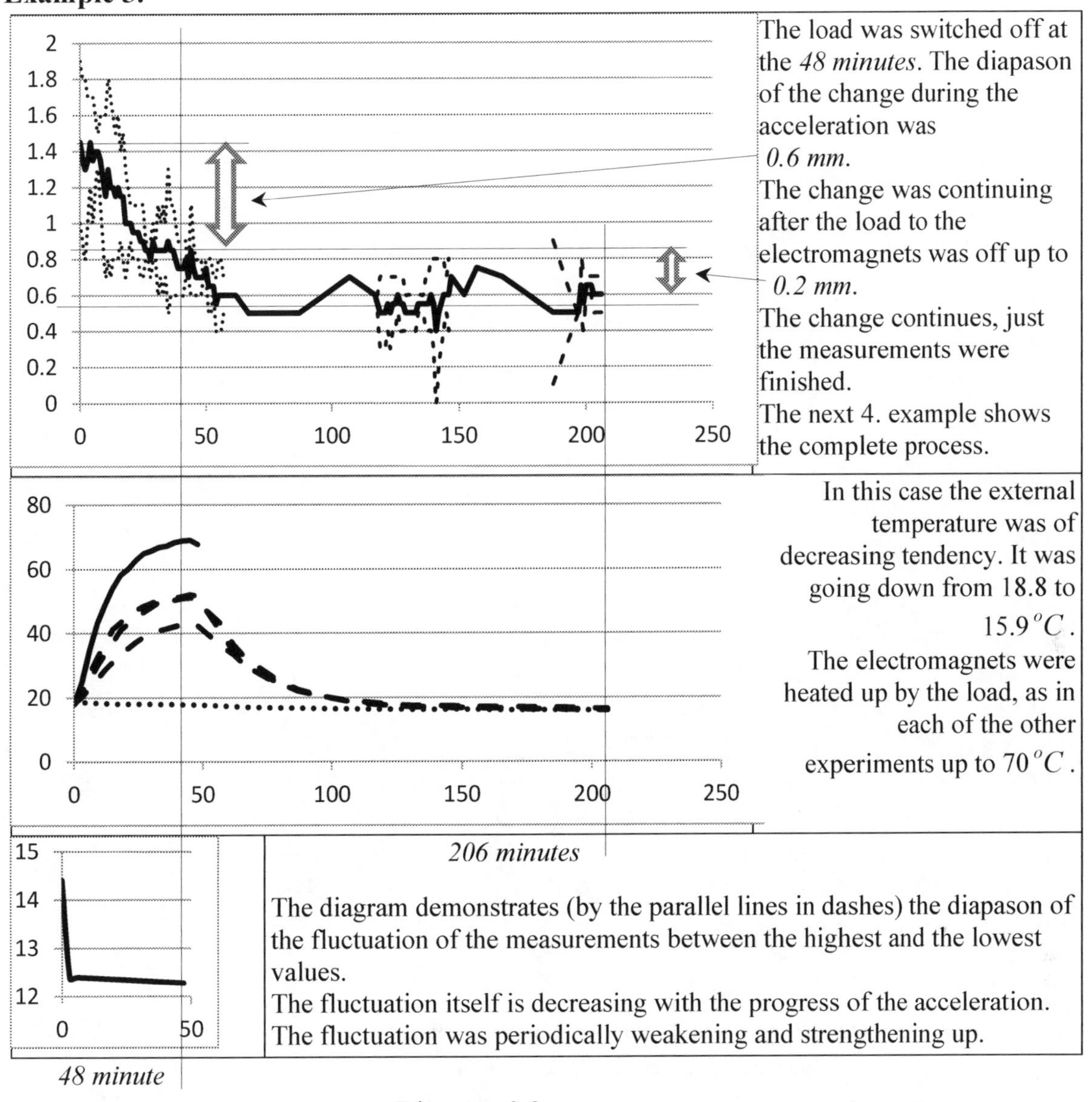

Diagram 9.3

Diag. 9.3

The heated up walls of the channel, made from *Cuprum* process, with higher quantum speed than the quantum speed of the *Hydrogen* process has their impact during the cooldown. The *IQ* drive of the *Cuprum* process of the wall reflects back all *Hydrogen* process internal impacts. There is no way the *Hydrogen* process inside could lose its increased status in similar tempo with the wall of the channel. The cooldown of the *Hydrogen* process this way is dictated by the cooldown of the *Cuprum* process and follows it. (The increased intensity characteristics of the *Cuprum* process do not influence the quantum impact of gravitation.)

The weight reduction stops, when the thermo-impact of the *Cuprum* process stops. The *Hydrogen* process comes back to its normal, not accelerated status within the channel with step by step growth, losing on its intensity.

This is presented by the next example of increased duration of the assessment.

Exp. 4

Example 4.

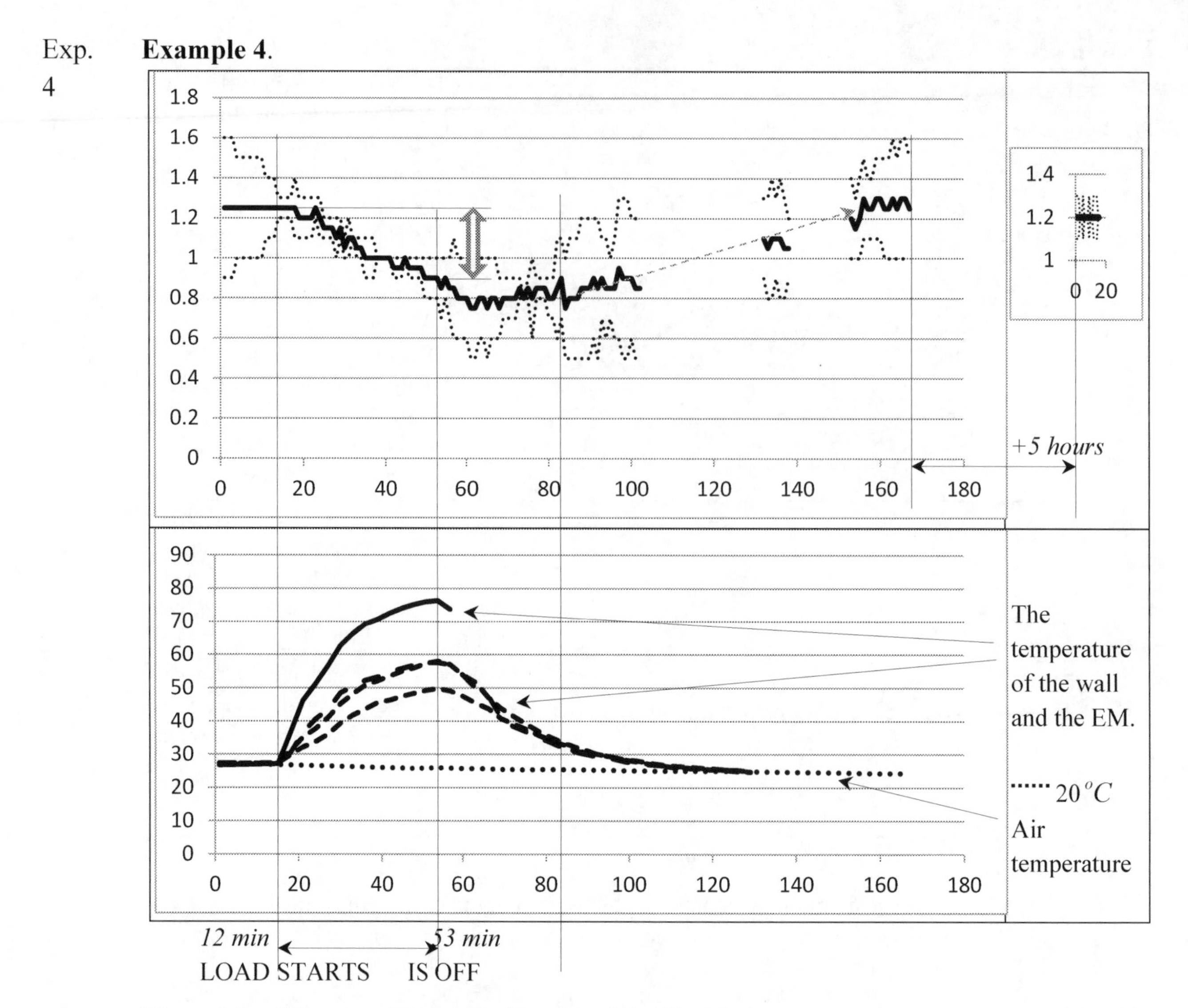

The weight of the system comes back to the original in 5 hours.

The diagram also shows the fluctuation of the weight values.

There were no measurements between the 102-132, the 152-167 minutes and in the final 5 hours of the rehabilitation. The data of the short measured periods between are well in line with the growth of the weight of the system.

Diag. 9.4

Diagram 9.4

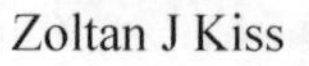

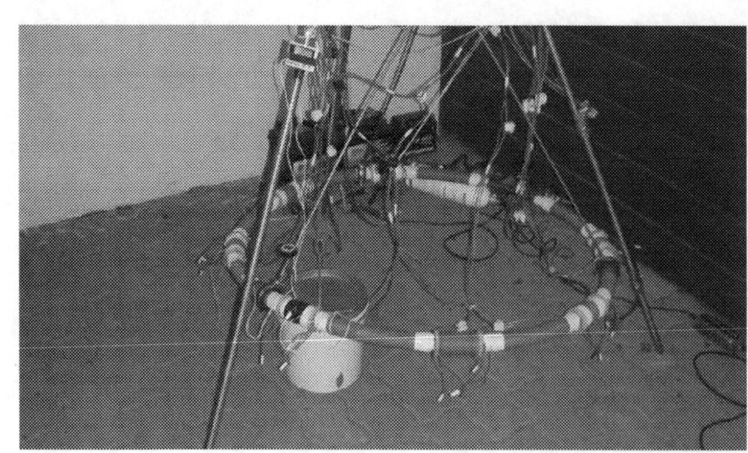

Picture 9.1 is about the segment of the construction of the accelerator: the channel with electromagnetic coils all around, the *Hydrogen* balloon connected, the 3 accumulators behind, the control weight balancing the channel and the measurements.

Pic. 9.1

Picture 9.1

9.3
The conclusion on the measurements

S. 9.3

On the basis of the extremely high number of measurement results, introduced here only through four examples, it can be stated that even this small scale experiment with this simple acceleration technique proves the existence of the conflict of the *Hydrogen* process in acceleration and of the quantum impact of *gravitation*.

The voltage of the load of the acceleration is decreasing in each case (as it is coming from an accumulators and even in the case of charging in parallel the accumulators, the demand is higher), having this way its certain impact on the internal dynamism of the intensity of the *Hydrogen* process. In the case the voltage of the induction to the electromagnets would be of constant value, the decreasing weight impact would also be more constant and stable.

If the channel of the acceleration would have its fix ends to the *Earth* surface, without compensating the change by its weight impact, and the load of the electromagnets would be of higher value, the conflict would cause measurable increase in the temperature of the *Hydrogen* process inside.

For benefiting from the acceleration of the *Hydrogen* process in industrial scale the *Hydrogen* process shall be speeded up close to the speed of the quantum communication (the speed of light) of the *Earth*. The source of the energy even in this case is not the *Hydrogen* process rather *gravitation*!

The acceleration of the *Hydrogen* process close to the speed of light results in conflict of infinite high intensity!
The conflict generates infinite high thermal energy.
The acceleration of the *Hydrogen* process means, the electron processes are impacting more number of quantum for the unit period of time.

If the normal impact, with reference to 9A1 was $e_H = \frac{dmc_E^2}{dt_E \varepsilon_H}\left(1-\sqrt{1-\frac{(c_E - i_E)^2}{c_E^2}}\right);$ 9D1

In the case of the acceleration, with reference to 9A3, it becomes:

$$e_{Hv} = \frac{dmc_E^2}{dt_E \varepsilon_H \sqrt{1-\frac{v^2}{c_E^2}}}\left(1-\sqrt{1-\frac{(c_E - i_E)^2}{c_E^2}}\right);$$ Ref. 9A3

With reference to 1A5: $m = f\left(\frac{n_H}{c_E^2}\right);$ and $mc_E^2 = f(n_H);$ | The mass is function of the quantum impact and the speed value of the quantum communication. Ref. 1A5

this way $$e_{Hv} = \frac{d[f(n_H)]}{dt_E \varepsilon_H \sqrt{1-\frac{v^2}{c_E^2}}}\left(1-\sqrt{1-\frac{(c_E - i_E)^2}{c_E^2}}\right);$$ 9D2

where neither dt_E, the time count, nor ε_H, the intensity coefficient of the *Hydrogen*

process can be subject to any change. v is the speed of the acceleration and c_E is the speed of the quantum communication of the *Hydrogen* process on the *Earth* surface. The result of the acceleration is the increase of n_H, the quantum impact of the electron process.
This increased quantum drive does not cause the increased intensity of the neutron collapse, as the neutron process of the *Hydrogen* process remains of infinite low intensity in any circumstances.

9D3 The increased number of the quantum impact is: $dn_{Hv} = \dfrac{dn_H}{\sqrt{1-\dfrac{v^2}{c_E^2}}}$.

(In the case of any other elementary process the increased intensity of the electron process would result in increased neutron collapse with the increase of the mass of the elementary process.)

The measured weight of the construction during the speeding up becomes less than its normal weight value at rest. The conflict between the increased quantum impact of the *Hydrogen* process and the quantum impact of gravitation is the one, which compensates the "lost" weight portion.

9D4
$$\Delta e = \frac{dmc_H^2}{dt_E \varepsilon_H}\left(1-\sqrt{1-\frac{(c_E-i_E)^2}{c_E^2}}\right)\left(\frac{1}{\sqrt{1-\left(v^2/c_E^2\right)}}-1\right);$$

The beauty of the case is, that while the conflict is developing between the *quantum impact of gravitation* and the *quantum impact* of the accelerated *Hydrogen* process, the result modifies the appearance (the weight), the *mechanical impact of gravitation.* In other words: the intensity relation of the conflict modifies the force impact of gravitation.

If we denote the "weight reduction impact" of the acceleration as function of the conflict of the acceleration of the *Hydrogen* process and the quantum impact of gravitation (*qig*) through:

9D5 $f(a_H \leftrightarrow qig)$

but at the same time the channel is fixed, the conflict is generating certain ΔG impact, independently is there any weight value to be measured or not. The main point here however is not about the weight reduction, rather the number of the impacted quantum the speeded up electron process of the *Hydrogen* process makes.

The increasing quantum impact of the *Hydrogen* process cannot be left without response from the side of gravitation. There is no way the increased quantum impact of the *Hydrogen* process withholds the quantum impact of gravitation or the continuity of the quantum impact of gravitation becomes damaged.
The higher the quantum impact of the acceleration is, the higher is the response from the side of gravitation: the higher is the conflict. This is not about the increase of the value of the background radiation, rather about the increase of the number of the responding anti-electron process *blue shift* impacts of the elementary evolution within the *Earth*.

In the case of a free to move accelerating channel, the conflict would either modify the weight value or would result in the vertical motion of the channel of the acceleration.
The closer the speed of the acceleration to the speed of light is, the higher is the value of the energy generation of the conflict – and higher the impact is.

Any weight reduction or vertical motion of the accelerating channel is the automatic consequence of the mechanical impact.

If the channel is free to motion, it is either reduces its weight, or the developing force impact is lifting it up.

$\Delta G = G - f(a_H \leftrightarrow qig)$; 9D6

if $f(a_H \leftrightarrow qig) > G$ the resulting "$-\Delta G$"corresponds to the lifting force impact;
If the accelerating channel is fixed, this automatic impact as force effect will be added to the conflict and increases the heat generation of the conflict.

$Q = f(a_H \leftrightarrow qig) + \Delta G$; the mechanical impact is added to the existing conflict. 9D7

The environmental impact of the acceleration at industrial scale is:

The intensity of the anti-neutron process at the inflexion is equal to the intensity of the neutron collapse. The intensities of the accumulation of the energy of the anti-proton processes are identical for all elementary processes (just the durations are variant).
Once, as consequence of the acceleration, the quantum impact of gravitation becomes of increased value, the expanding rate of the anti-neutron processes of the elementary processes of the elementary evolution becomes of higher proportions. (As there is no way any modification in the quantum impact of gravitation happens, while the demand in anti-electron process impacts becomes increased.)
The anti-proton processes require certain number of acting anti-electron process drives. If the generating surplus expires, it speeds up the intensity of the neutron process, which in fact speeds up the whole chain of the proton expansion – neutron collapse processes.
The environmental impact of the acceleration of the *Hydrogen* process in the elementary evolution of the *Earth* is equivalent to mining out minerals. Mining out minerals reduces the generation of the anti-electron processes, the drives of the anti-proton process collapse the same way as the demand of the conflict with the speeded up *Hydrogen* process.

A

Attachment

Abstract from the book: *The Energy Balance of Relativity*
published in 2007, with actual explanations and small corrections

Review of Time and Space Coordinate Relations of the Special Theory and the Foundation of the General Theory of Relativity

Einstein's papers, "On the Electrodynamics of Moving Bodies" (A) of 1905, and
"On Influence of Gravitation on the Propagation of Light" (B) of 1911,
are the subjects of the review. The chapter also comments
"The Foundation of the General Theory of Relativity" (C), the article of 1916.
For simplicity, the original characters and definitions were used.

S. A.10.1

A.10.1
Review of the theory on moving rigid bodies and moving clocks

We are taking the systems of reference as described in Chapter 4 of *Einstein's* paper, 'A'.

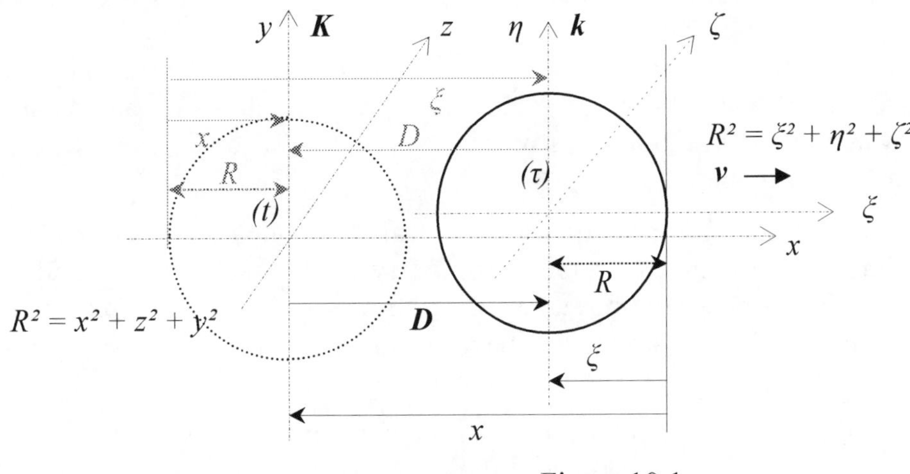

Figure 10.1

Fig. 10.1

There are two systems of references in Figure 10.1 above:

- *K:* with coordinates *x, y* and *z*, with time measurement *t*.
- *k:* with coordinated *ξ, η* and *ζ*, with time measurement *τ*.

  There is a rigid sphere, radius of *R*, positioned in the centre of *k*.

What the description of *k* in motion with $v = const$ relative to the stationary *K* is? This is the original example, *Einstein* has examined
The common position of the origins of *K* and *k* are $x = 0$ and $\xi = 0$ at a measured time moment of $t = \tau = 0$ accordingly.
The systems are in motion with $v = const$ relative to each other.

For the better presentation of the case in Figure 1.1, axis *ξ* is shifted slightly upward, but axes *x* and *ξ* are supposed to be coincided.

$$\text{Obviously: } \Delta t = t - 0 = t \quad \text{and} \quad \Delta\tau = \tau - 0 = \tau \qquad \text{10A1}$$

The examination was made and established, that the time count in the system of reference in motion (k), relative to the stationary other (K) is slower:

10A2
$$\Delta t_{E-motion} = \Delta t_{E-stat}\sqrt{1-\frac{v^2}{c^2}};$$

and the lengths of the space coordinates of the sphere in motion are measured as shorter:

10A3
$$\Delta l_{E-motion} = \Delta l_{E-stat}\sqrt{1-\frac{v^2}{c^2}}.$$

On the basis of the findings *Einstein* has stated, that in systems in motion time ticks slower and space coordinates become shortened.

The purpose of the following assessment is to prove: this formulation is wrong, as the *Einsteinian* examination was incomplete.

The reason of the incompleteness is the fact that the examination was made only from the point of view of K, the stationary system. The case must be examined equally from the point of view of both systems of reference – which as fact missing from *Einstein* study:

- the case, when k is in motion with speed $v = const$ relative to the supposed to be stationary K, and also the opposite version,
- the case, when k is the system of reference supposed to be the stationary one and K is in motion with $v = const$ relative to k.

Whatever system from the two is taken for being in motion with $v = const$, both systems are equally in motion relative to the other. Their congruent position is over and there is no way to define which one is in motion and which one remains at rest. This is the reason, both versions shall be examined.

Current examination is looking for the description of the time duration of the motion and the space coordinates of the sphere of the supposed system in motion via the identical parameters of the other, supposed to be stationary system. Speed value $v = const$ remains de facto one and the same in both scenarios. The transformation equations are equally valid for both cases. For simplicity, only coordinates x and ξ were examined.

The time and the space coordinates in line with *Lorenz*-transformation:

10A4
10A5
10A6
10A7

if <u>*k is in motion*</u> relative to the supposed K at rest	if <u>*K is in motion*</u> relative to the supposed to be stationary k
$\tau = \frac{t-(vx/c^2)}{\sqrt{1-(v^2/c^2)}}$; and $\xi = \frac{x-vt}{\sqrt{1-(v^2/c^2)}}$;	$t = \frac{\tau-(v\xi/c^2)}{\sqrt{1-(v^2/c^2)}}$; and $x = \frac{\xi-v\tau}{\sqrt{1-(v^2/c^2)}}$

$$x = ct \text{ and } \xi = c\tau$$

*

For the correct statement of the relations it is important to clarify, what the source of the energy, necessary for driving the motion is? Is this an energy, coming from an external source, or this is the energy, utilised of the system in motion?

S.A.10.1 1

A.10.1.1 <u>*k is in motion relative to the stationary K*</u>

10B1
10B2

With reference to Figure 10.1/a on the next page and from the supposed scenario follows, that while system k makes distance $D = v \cdot t$ within the stationary K, the light signal with the information about the motion makes distance $x = c \cdot t$ from the fares spot of the radius of the sphere to the centre of K. (The continuous lines.)

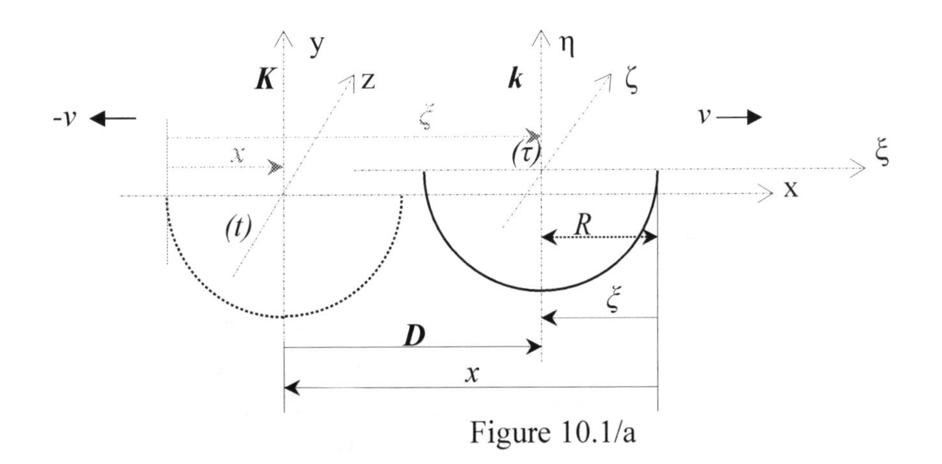

Figure 10.1/a

Fig 10.1/a

Ref. 10A6

With reference to 10A6, the transformation equation, the distance within system k in motion corresponds to: $$\xi = \frac{x - vt}{\sqrt{1-(v^2/c^2)}}$$ 10B3

This way, the distance, the light signal from the fares spot of the sphere of system k in motion to the centre of the stationary K, in k makes corresponds to: $$\xi = \frac{ct - vt}{\sqrt{1-(v^2/c^2)}};$$ 10B4

There is also a light signal with information, arriving from the fares spot of the sphere of the stationary system K to the centre of system k in motion, making a distance of $\xi = c \cdot \tau$, while system k makes distance D. 10B5

10B1, 10B2, 10B3, 10B4 and 10B5 give: $$c\tau = \frac{ct - vt}{\sqrt{1-(v^2/c^2)}};$$ 10B6

The measured time while the event happens, in system K is t, and in system k is τ. The relation of the two is: $$\tau = t\frac{1-(v/c)}{\sqrt{1-(v^2/c^2)}};$$ 10B7

A.10.1.2. *While system k is still in motion indeed, K is seen being in motion from the point of view of k, supposed to be stationary relative to its own system of reference*

S. A.10.1 2

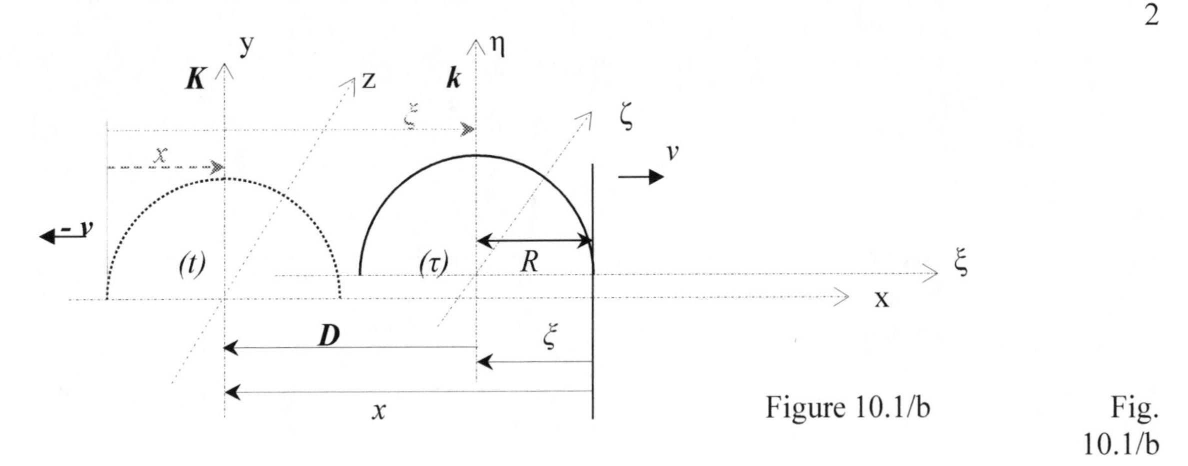

Figure 10.1/b

Fig. 10.1/b

Why this version shall also be examined?
The answer is simple: as the event could happen in two ways, its interpretation must reflect both versions.

10C1 If the motion of system *k* has been taken as $+v = const$ relative to *K*,
10C2 than the supposed in this time motion of system *K* is $-v = const$ relative to *k*, supposed to be at rest; and the distance *K* "makes" within the system of reference *k* is $D = (-v) \cdot t$, while the distance of the light signal, measured within *k* is: $\xi = c \cdot \tau$.

(Still the continuous lines to follow in Figure 1.1/b on the previous page.)

The measurement has been similarly made from the centre of system *K* and the subject of the measurement is the same way the fares coordinate ξ of system *k*.
While the circumstances of the measurement were changed, the event has remained the same and the result cannot be different!

10C3 Following the steps of the previous example: $x = c \cdot t = \dfrac{\xi - (-v)\tau}{\sqrt{1-(v^2/c^2)}} = \dfrac{c\tau + v\tau}{\sqrt{1-(v^2/c^2)}}$;

10C4 the duration of the event from 10C3 above is: $t = \dfrac{\tau + (v/c)\tau}{\sqrt{1-(v^2/c^2)}} = \tau \dfrac{1+(v/c)}{\sqrt{1-(v^2/c^2)}}$;

Even these preliminary results prove, the reciprocal examination was certainly necessary: The result in 10B6 and 10B7 are with minus, here above in 10C3 and 10C4 are with plus.

10B7 and 10C4 give the following result:

10D1
10D2

$$\frac{\tau_{(rel.m)}}{t} = \frac{1-(v/c)}{\sqrt{1-(v^2/c^2)}}; \qquad \frac{t_{(rel.m)}}{\tau} = \frac{1+(v/c)}{\sqrt{1-(v^2/c^2)}}$$

The durations, measured within the systems in motion have index *(rel.m)* in the equations.

S. A.10.1 3 *A.10.1.3. <u>Only the parallel assessment of all variants can give valid result!</u>*

For the completeness of the case and for the easy assessment of the measured data, system *K* now is proposed to be in motion within system *k*.

(And the lines in dash now shall be followed.)

While physical parameters of the case, the speed of the motion and the distance are the same, the interpretations of the event are again of two scenarios.

The measured data are written into the equation of the *Lorentz*-transformation, the direction of the motion indicated by *D*.
For the first variant (Figure 10.1/a in dash) the result is:

10D3 $\dfrac{\tau_{(rel.m)}}{t} = \dfrac{1+(v/c)}{\sqrt{1-(v^2/c^2)}}$; the direction of the motion is negative, (as it was taken in the second scenario), because *K* moves relative to *k* in negative direction.

10D4 as: $\xi = c \cdot \tau = \dfrac{x - (-v)t}{\sqrt{1-(v^2/c^2)}} = \dfrac{ct + vt}{\sqrt{1-(v^2/c^2)}}$; and $\tau = \dfrac{t + (v/c)t}{\sqrt{1-(v^2/c^2)}} = t \dfrac{1+(v/c)}{\sqrt{1-(v^2/c^2)}}$;

For the second variant (Figure 1.1/b in dash) the result is:

10D5 $\dfrac{t_{(rel.m)}}{\tau} = \dfrac{1-(v/c)}{\sqrt{1-(v^2/c^2)}}$; since the motion here is away from the point of the measurement.

10D6 as: $x = c \cdot t = \dfrac{\xi - v\tau}{\sqrt{1-(v^2/c^2)}} = \dfrac{c\tau - v\tau}{\sqrt{1-(v^2/c^2)}}$; and $t = \tau \dfrac{1-(v/c)}{\sqrt{1-(v^2/c^2)}}$;

There are a couple of important points to be noted before coming to the expression of the time relations:

- the relative motion of two systems is a single event, but with the necessary parallel assessments;
- k is in motion relative to K; and with identical impotence: K is in motion relative to k;
- for positioning the event in space and time the correct way and for the precise definition of the relation of the systems – the two scenarios of both variants shall be assessed.

The selection of the "minus" and the "plus" direction is free, just with their consequent use in the *Lorentz-transformation.*

The time relation of the motion can be expressed from 10D1, 10D2, 10D3 and 10D5:
For the examined scenarios for both variants give:

$$\frac{\tau_{(rel.m)}}{t} + \frac{t_{(rel.m)}}{\tau} = \frac{2}{\sqrt{1-\left(v^2/c^2\right)}}; \qquad \text{10D7}$$

and for both systems in motion separately give:

$$2\frac{\tau_{(rel.m)}}{t} = \frac{2}{\sqrt{1-\left(v^2/c^2\right)}}; \quad \text{and} \quad 2\frac{t_{(rel.m)}}{\tau} = \frac{2}{\sqrt{1-\left(v^2/c^2\right)}}; \qquad \text{10D8} \quad \text{10D9}$$

10D7, 10D8 and 10D9 prove:
in the case of two systems in motion relative to each other with $v = const$ the time relations are identical:

$$\frac{\tau_{(rel.m)}}{t} = \frac{t_{(rel.m)}}{\tau} = \frac{1}{\sqrt{1-\frac{v^2}{c^2}}} \qquad \text{10D10}$$

10D10 means, the time count is speeded up in the system in motion relative to the system at rest:

$$\frac{\Delta\tau_{(rel.m)}}{\Delta t} = \frac{\Delta t_{(rel.m)}}{\Delta\tau} = \frac{1}{\sqrt{1-\frac{v^2}{c^2}}}; \quad \text{or different way:} \quad \frac{d\tau_{(rel.m)}}{dt_{(rel.rest)}} = \frac{dt_{(rel.m)}}{d\tau_{(rel.rest)}} = \frac{1}{\sqrt{1-\frac{v^2}{c^2}}}; \qquad \text{10D11}$$

Formulas in 10D11 above are the *greatest proofs of relativity*. They address the time relation, *from the point of view of the system of reference supposed to be in motion* relative to the other, supposed to be the stationary one

- *the reciprocal character of the motion with speed* $v = const$;
- *the equality of the systems in relative motion with relative speed* $v = const$.

Meaning: the time flow *is faster* within the system of reference *in relative motion*!
The expression in 10D11 above is reciprocal to the one, given by *Einstein* in his classical time relation formula!

With reference to * on page 2 and to the above, it is important to note that the time relation between systems of reference in motion and rest depend on the fact, where the energy, the source of the motion is coming from. Is the system of reference in motion for the count of its own energy, like it is presented in 10D11, or the energy (work) of the motion is coming from an external source?

Ref * page 2

The time count characterises the intensity of the event in the system of reference.
There is therefore a huge difference, is the system in motion using its own energy as drive of the motion, or the energy source of the motion is coming from an external source.
The example of the difference is given in the following paragraph.

S. A.10.1 4

A.10.1.4. Further clarification of the time relations, taking into a count the drive of the motion

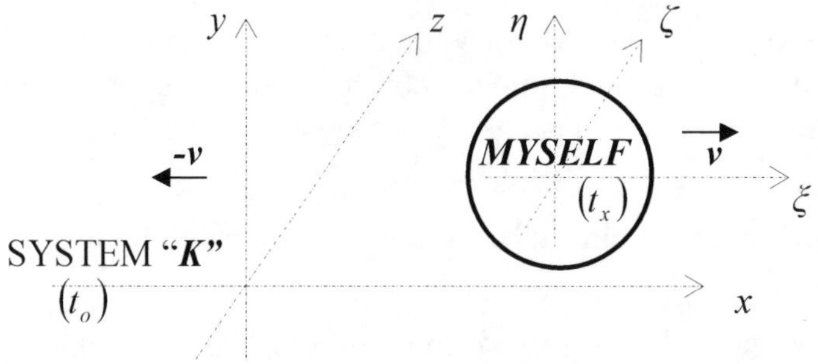

Fig. 10.2

Figure 10.2

The time count of my system of reference, if I am the drive of my own motion in system of reference K, with reference to 10D11 is: $t_x = \dfrac{t_o}{\sqrt{1-\dfrac{v^2}{c^2}}}$;

10X1

(as I am the one in motion relative to the stationary K)

and this is the description of the change of my intensity (energy source) within the system of reference of rest, K: $\Delta e_x = \dfrac{dmc^2}{dt_o}\left(1-\sqrt{1-\dfrac{v^2}{c^2}}\right)$;

10X2

If I have been externally driven by a source in system K, meaning, I do not use any own energy, but still in motion, this is identical to the case, K is in motion relative to my stationary system. (I do not have any impact to my motion.)

This means, the time count of K with reference to 10D11 – as system K is in motion relative to my stationary system of reference – is $t_o = \dfrac{t_x}{\sqrt{1-\dfrac{v^2}{c^2}}}$;

10X3

And the time count in my own system is: $t_x = t_o\sqrt{1-\dfrac{v^2}{c^2}}$

10X4

And the energy intensity of my existence, using the parameters of the system of reference of K becomes: $\Delta e_x = \dfrac{dmc^2}{dt_o}\left(1-\dfrac{1}{\sqrt{1-\left(v^2/c^2\right)}}\right)$;

10X5

(actually negative, since receiving energy)

The time count of my internal system in motion (for the count of an external drive, located in system K) is important, since this is the one characterising my intensity (energy) capacity. And with reference to 1X4, by the increase of the speed of my motion, my time count is shortening (slowing down)!

Ref. 10X4

If the time count at rest was t_{xo} and it becomes t_{xv} in motion, the change of the intensity (energy) capacity of my motion is: $\Delta e_x = \dfrac{dmc^2}{dt_{xo}} - \dfrac{dmc^2}{dt_{xv}}$;

10X6

I cannot be obviously in motion relative to myself, therefore the time count of rest in this case shall correspond to the time count of the system of reference of my existence. The one in fact at rest: $t_{xo} = t_o$

And the value of t_{xv} is coming from the formulas of the previous explanations.

If I am the drive of my own motion	If my motion is driven from external source:	
$t_{xv}=\dfrac{t_{xo}=t_o}{\sqrt{1-\dfrac{v^2}{c^2}}};$	$t_{xo}=t_o=\dfrac{t_{xv}}{\sqrt{1-\dfrac{v^2}{c^2}}};$ and $t_{xv}=t_{xo}\sqrt{1-\dfrac{v^2}{c^2}};$	10X7 10X8
meaning: $\Delta e_x=\dfrac{dmc^2}{dt_{xo}}\left(1-\sqrt{1-\dfrac{v^2}{c^2}}\right);$	and this way: $\Delta e_x=\dfrac{dmc^2}{t_{xo}}\left(1-\dfrac{1}{\sqrt{1-\left(v^2/c^2\right)}}\right);$	10X9 10X10

A.10.1.5. The relation of space-coordinates

S. A.10.1 5

The used for the time relations methodology, for the space-coordinates gives the following results:

The space coordinates for the two sensations of both cases are::

$$\xi_{(rel.m)}+x_{(rel.m)}=\frac{x-vt}{\sqrt{1-\left(v^2/c^2\right)}}+\frac{\xi+v\tau}{\sqrt{1-\left(v^2/c^2\right)}}=\frac{x}{\sqrt{1-\left(v^2/c^2\right)}}+\frac{\xi}{\sqrt{1-\left(v^2/c^2\right)}};$$

Ref. 10B6 10C4 10D4 10D6 10E1

And separately for the two systems:

- for system of reference k :

$$\xi_{(rel.m)}=c\tau=\frac{ct-vt}{\sqrt{1-(v^2/c^2)}};\ \text{and}\ \ \xi_{(rel.m)}=c\tau=\frac{ct+vt}{\sqrt{1-\left(v^2/c^2\right)}};\qquad \xi_{(rel.m)}=\frac{x}{\sqrt{1-(v^2/c^2)}};$$

Ref. 10B6 10D4 10E2

- for system of reference K:

$$x_{(rel.m)}=c\cdot t=\frac{c\tau+v\tau}{\sqrt{1-\left(v^2/c^2\right)}};\ \text{and}\ \ x_{(rel.m)}=c\cdot t=\frac{c\tau-v\tau}{\sqrt{1-(v^2/c^2)}};\qquad x_{(rel.m)}=\frac{\xi}{\sqrt{1-\left(v^2/c^2\right)}};$$

Ref. 10C3 10D6 10E3

$\xi_{(rel.m)}=\xi$ and $x_{(rel.m)}=x$ only in the case, if $v = 0$.

10E4

10E2 and 10E3 can also be written in other formats as well:

$$\frac{d\xi_{(rel.m)}}{dx_{(nyugalom)}}=\frac{1}{\sqrt{1-\dfrac{v^2}{c^2}}};\qquad \frac{dx_{(rel.m)}}{d\xi_{(nyugalom)}}=\frac{1}{\sqrt{1-\dfrac{v^2}{c^2}}};$$

10E5

10E5 gives that valid distance, the system of reference in motion makes relative to the stationary system of reference. This means, contrary to the *Einsteinian* definition in 10A3 – the distance, the system of reference in motion makes is longer than it is measured with the system of reference at rest.

Ref. 10A3

In line with this, the distance k in relative motion to K makes, measured within the stationary system of K is (Figure 10.1/a): $D_K=v\cdot t_K$

The same distance, but measured within k, the system in motion : $D_k=v\tau=v\dfrac{t_K}{\sqrt{1-\left(v^2/c^2\right)}}$; and $D_k>D_K$

Ref. 10B2 10E6

In the case, system K is in motion relative to k, the distance K makes within k, but measured in k is (Figure 10.1/b): $D_k=v\cdot\tau_k$

And the same distance, but measured in K in motion is: $D_K=vt=v\dfrac{\tau_k}{\sqrt{1-\left(v^2/c^2\right)}}$; and $D_K>D_k$

Ref. 10C3 10E7

10E8 The speed of the motion is equal from the point of view of both systems of reference: $\frac{d\xi_{(rel.m)}}{d\tau_{(rel.m)}} = \frac{dx_{(rel.m)}}{dt_{(rel.m)}} = \frac{d\xi_{(nyugt)}}{d\tau_{(nyug)}} = \frac{dx_{(nyug)}}{dt_{(nyug)}} = v$

The dimensions of subjects at rest within the systems of reference in motion are without change. They remain of the same size!

S. A.10.2

A.10.2
Concerns about *Einstein's* formula

1./
The valid relations of time and space-coordinates between systems of reference in motion and at rest are reciprocal to the formulas proposed by Einstein!

2./
Einstein states in his papers that *lengths* of measuring rods and *rates* of clocks are variable in systems of reference in motion.

But there is no change in the length of measuring rods and in the rate of clocks. Standardised clocks and measuring rods are of one and the same rate and size in any systems of reference.

If the measurements are of different values, it is not because of any change in length-units or clock-rates. These are real measurements in frequencies (time) and wave lengths (distance) as a result of the modified time flow (intensity) within the system of reference in motion.

3./
Can the *physical distance* between systems of reference in motion and at rest be different?

No!

The *de facto* distance is one and the same, but its measured value depends on the motion of the system of reference where the measurement (*the event*) takes place. The *variable* is the *duration of events*, resulting in *variable time flow* within systems of reference distinct in motion.

Any measurement, even of static dimensions is an event. And any event can be characterised by the time necessary for the measurement, and so the "intensity" of the event (measurement) is:

10F1
$$quotient = \frac{l}{\Delta t} = \frac{length-(light-motion)-of-subject-to-be-measured}{duration-of-the-measurement-(event)};$$

The intensity of length measurement (while the light signal passes the length)

in the stationary system of reference is:	while in the system of reference in motion:
$\varepsilon_o = \frac{dl_o}{dt_o};$	$\varepsilon_v = \frac{dl_v}{dt_v};$

10F2
10F3

And the measured distances are different indeed:

10F4
$$\Delta l_o = \varepsilon_o \cdot \Delta t_o; \quad \text{és} \quad \Delta l_v = \varepsilon_v \cdot \Delta t_v \quad \text{and} \quad \Delta l_o \neq \Delta l_v$$

While the time flow is variable, static geometrical dimensions in systems of reference in motion are *invariant*!

A.10.3
Review of the statement on *Gravitation* of Energy

S. A.10.3

Einstein compares the effects of gravitation and acceleration in Chapter 2 of 'B'.
In fact, this is not a real '*what-is-the-reason*' type comparison, where we can find *pros* and *contras* on gravitation and acceleration concerning the origin of the *increase* of the frequency. Rather this is an *a priori* statement on the reason, as if it would only be the effect of gravitation, and the explanations are given accordingly.

We will prove that acceleration has the (same) characteristics *Einstein* has attributed to *gravitation*.

We are taking the same example, the same figure and similar markings as in Chapter 2 of 'B' and will show that there should be *nothing* that gives priority to gravitation:
A system of reference, denoted as S_a is in acceleration with $a = const$ (where $a = -g$) within a stationary system of reference, denoted as S_o without gravitational field and with homogenous radiation of frequency f_o within it.

S_a is accelerating upward from position S_{1a} to position S_{2a}.

The time measurements within S_o and S_a are t_o and t_a accordingly.

We are looking for the frequency of the radiation, measured in collision with S_a, at the surface of S_a.

[For better orientation, the parameters of the systems of reference in acceleration and at rest, taken from *Einstein's* example in 'B', are additionally denoted through *'a'* and *'o'* accordingly.]

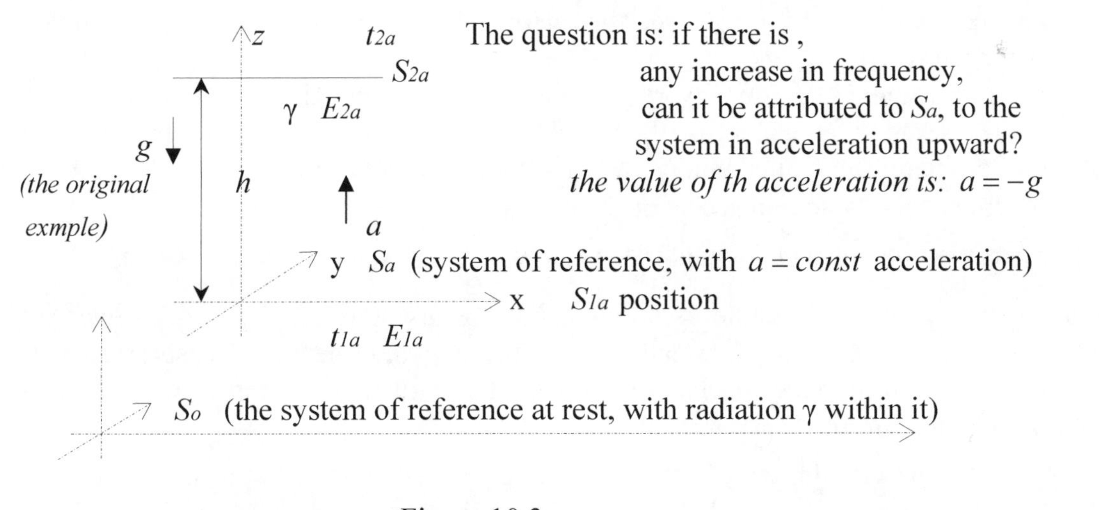

Figure 10.3

Fig. 10.3

The example supposes that the collision will be taking place at position S_2 at time moment: $t_o = \dfrac{h}{c}$ (10G1)

and the velocity of S_a at this moment will be: $v = a\dfrac{h}{c}$ (10G2)

(a and h, both measured within S_o)

We apply the transformation equations to determine the values of the frequency and the wavelength of the radiation for the case with *no gravitational* field:

We measure and compare the frequencies of the impacts (named "photons" in *Einstein* example) of energy E_{1a} at position *Sa* at $v=0$, the start of the motion of *S1a* upward to position *S2* and at the moment of the arrival of *S1a* at position *S2* at the speed of $v=a\frac{h}{c}$.

10G3 Ref. 10D11 10G4

At a certain time moment, while *S1a* makes distance *h* and the actual speed is $v=-g\frac{h}{c}$, the time relation of the two systems of reference is: $t_a=\frac{t_o}{\sqrt{1-(v^2/c^2)}}$

We use the transformation equations to characterise the frequencies and the wavelengths of the quantum impact (the photons in the original), radiated within *So* to *Sa*.

10G5 Ref. 10A4 10A5 10H1

From 10A4 and 10A5, and taking account that: $-(-v)=-\left(-g\frac{h}{c}\right)=g\frac{h}{c}=v$;

$$\frac{1}{f_a}=\frac{\frac{1}{f_o}+\frac{v\lambda_o}{c^2}}{\sqrt{1-\frac{v^2}{c^2}}};\qquad \left(f=\frac{1}{t};f=\frac{c}{\lambda}\right)\qquad \lambda_a=\frac{\lambda_o+\frac{v}{f_o}}{\sqrt{1-\frac{v^2}{c^2}}};$$

10H2

substituting ($c=f\cdot\lambda$) in 10G1 and 10G2, both give $\frac{1}{f_a}=\frac{1}{f_o}\frac{1+(v/c)}{\sqrt{1-(v^2/c^2)}}$

10H3

which gives $\frac{1}{f_a}=\frac{1}{f_o}\sqrt{\frac{1+(v/c)}{1-(v/c)}}$; equal to $f_a=f_o\sqrt{\frac{1-(v/c)}{1+(v/c)}}$

The measurement takes place upon the impact (the collision of photons in *Einstein* example) at the surface of *Sa*. Let us see the energy balance *before* and *after* the impact (collision) at position *S2a*.
Meaning: we measure the frequency of the impact at two positions of the system of reference in acceleration: at *S1a*, before the impact (called collision), on a stationary system and at *S2a*, after the impact (called collision), on a system in motion.
E_{1a} is the energy of the system in acceleration at position *S1a*.
E_o is the energy of the impact (of the photons called in *Einstein* example) of the radiation.

For making the equation as simple as possible, we are taking *S1a* and *S2a* as unique and discrete points and distance *h,* as small as possible. No intermediate measurements of frequency between them take place. The impact (the flow of photons by *Einstein*) is continuous, and their original frequency is f_o.
The energy of the system of reference in acceleration before the impact at position *S1a*:

10H4 $E_{1a}=f_{1a}\cdot H$; at *S1a*: $E_{1a}=E_o$ and $f_{1a}=f_o$

The energy of the radiation at position *S2a*: $E_{2a}=f_{2a}\cdot H$ *H* is the *Planck-constant*
After the impact (collision as *Einstein* names it) at position *S2a*:

10H5 $$E_{2a}=E_{1a}+\Delta E_a \qquad [\text{ since } m\cdot a\cdot h=\frac{E_{1a}}{c^2}ah]$$

10H6 The energy of the radiation within the stationary system is: $E_o=f_o\cdot H$

The change of the kinetic energy between *S2a* and *S1a* is:

10H7 $$\Delta E_a=E_{2a}-E_{1a}=(f_{2a}-f_{1a})H;\quad \text{or}\quad \Delta E_a=(f_{2a}-f_o)H$$

Depending on the resulting value of the f_{2a},

$|f_{2a}| < |f_o|$ means: $(f_{2a} - f_o)H = -\Delta E_a$ $\rightarrow$ the increase of the frequency of the radiation results in *loss of kinetic energy*. The inert body or system of reference in acceleration *gives off* ΔE_a energy in form of radiation.

$|f_{2a}| > |f_o|$ obviously means the opposite: $(f_{2a} - f_o)H = \Delta E_a$ $\rightarrow$ the radiation *gives off* ΔE_a energy to the inert body or system of reference in motion.

With 10H5, 10H6 and 10H7 we have just come to the same conclusion for acceleration, as *Einstein* has made with his *a priori* statement on the example of radiation within gravitational field: An inert body or system of reference in acceleration *results in a shift of the frequency* of electromagnetic waves in collision.
10H8 and 10H9 below are corresponding to *Einstein's* equations

$$\frac{f_{2a} - f_o}{ah} H = -\frac{\Delta E_a}{c^2}; \qquad \frac{f_{2a} - f_o}{ah} H = \frac{\Delta E_a}{c^2} \qquad \text{10H8} \quad \text{10H9}$$

A.10.3.1. Coefficient of the energy transfer S. A.10.3 1

There is a *disharmony*, which must be resolved, between the values of frequencies and wavelengths deduced from the *transformation equations,* and *diverted directly* from the time relations of relative motion.

With reference to 10H3, the transformation equation for the *frequency* relations gives $f_a = f_o \sqrt{\frac{1-(v/c)}{1+(v/c)}}$ Ref. 10H3 10H10

With reference to 10D11: $t_a = \frac{t_o}{\sqrt{1-(v^2/c^2)}}$, Ref. 10D11

the *value of the frequency* directly diverted from the relativistic time formula is:

$f_{at} = f_o \sqrt{1 - \frac{v^2}{c^2}}$; since $t_{at} = \frac{1}{f_{at}}$; ('$t$' in the index denotes it comes for the time formula) 10I1

We manage the difference by a coefficient, denoted δ_f:

$$\delta_f \cdot f_a = f_{at}; \text{ and } \delta_f \cdot f_o \sqrt{\frac{1-(v/c)}{1+(v/c)}} = f_o \sqrt{1 - \frac{v^2}{c^2}}; \quad \text{and} \quad \delta_f \sqrt{\frac{1 - \frac{v}{c}}{1 + \frac{v}{c}}} = \sqrt{1 - \frac{v^2}{c^2}} \qquad \text{10I2}$$

The value of δ_f depends obviously on which part of the equation in 10I1 is to be adjusted.

Therefore, the value is either $\delta_f = 1 + \frac{v}{c}$; or $\delta_f = \frac{1}{1 + \frac{v}{c}}$; 10I3 10I4

If we take a closer look at 10I3 we recognize that this is the coefficient *Einstein* suggests one makes to correct the velocity of light, because of the presumed gravitational effect:

$$\delta_f = 1 + g\frac{h}{c^2}; \quad \text{or} \quad \delta_f = 1 + \frac{\Phi}{c^2}; \qquad \text{10I5} \quad \text{10I6}$$

What are these "coefficients to correct" for?

The time relations do not need energy considerations and, furthermore, those are reciprocal and independent of the direction of the motion. The motion itself and its direction, however, have real physical impact on the frequencies and wavelengths. Therefore, in frequency and wavelength relations the energy component is important.

Ref 10D11 10I1

The transformation equations initiate well the energy relations. For keeping the frequency in balance with the time relations, reference to 10D11 and 10I1, the radiation needs energy from the other participant of the collision, the inert body or system of reference in motion. If the energy is granted, the frequency of the radiation becomes level with the frequency, diverted directly from the relativistic time formula.

Coefficient δ_f is dedicated to the energy (frequency) correction.

10I7 10I8

With the correction by δ_f: $f_{1at} = \delta_f f_{1a} = f_{1a}\left(1+\frac{v}{c}\right)$; and $f_{1a}\left(1+\frac{v}{c}\right) = f_{2o}\sqrt{1-\frac{v^2}{c^2}}$;

10I7 and 10I8 have nothing to do with *gravitation*!
This is the way the harmony can be assured between the *Lorentz-transformation* and the equation of relativity deducted in correct, two scenarios way!
This is certainly necessary, since the time formula is not taking into account any energy balance considerations.

S. A.10.3 2

A.10.3.2. *The energy transfer and Doppler's principle*

Ref 10H3 10K1

We are taking 10H3 again $\frac{f_{1a}}{f_{2o}} = \sqrt{\frac{1-\frac{v}{c}}{1+\frac{v}{c}}}$; and write it in form of $\frac{f_{1a}}{f_{2o}} = A$

and looking for the values for A

Ref. 10H7 10K2

In general $\frac{dE_a}{df} = H$, the 'Plank-constant'. from 1H7: $\Delta E_a = (f_{2a} - f_o)H = \Delta f_a H$

Using the relation in 10K2 there are three cases to distinguish,
(marking f_{2a}, just through a more general f_a = for the system in motion):

10K3 ► case $v = 0$: $\Delta E_a = 0$ ► $\Delta f_a = 0$ ► $\frac{df_o}{df_a} = 1$ ► $\frac{f_o}{f_a} = 1$ ► $\sqrt{\frac{1-\frac{v}{c}}{1+\frac{v}{c}}} = 1$ ► $f_{1a} = f_o$

► case in 10H9: radiation *gives off* ΔE_a energy:

10K4 $\Delta E_a > 0$ ► $\Delta f_a > 0$ ► $\frac{df_o}{df_a} < 1$ ► $\frac{f_o}{f_a} = A$ ► $\sqrt{\frac{1-\frac{v}{c}}{1+\frac{v}{c}}} = A$ ► $f_o = f_a\sqrt{\frac{1-\frac{v}{c}}{1+\frac{v}{c}}}$

(in 10G5 $f_o < f_{2a}$ indeed!)

► case in 1H8: system of reference in acceleration *gives off* ΔE_{1a} energy

10K5 $\Delta E_a < 0$ ► $\Delta f_a < 0$ ► $\frac{df_o}{df_a} > 1$ ► $\frac{f_o}{f_a} = \frac{1}{A}$ ► $\sqrt{\frac{1-\frac{v}{c}}{1+\frac{v}{c}}} = \frac{1}{A}$ ► $f_o = f_a\sqrt{\frac{1+\frac{v}{c}}{1-\frac{v}{c}}}$

10K5 describes the usual form of the Doppler's formula for collision from in front.

But the inert body or system of reference of the collision must have and give off the energy, which ensures the increase of the frequency!
Meaning: The resource of the frequency increase is the kinetic energy of the inert body or system of reference in motion with $a=-g$.

10K2-10K5 give the complete energy characterization of the shift of the frequency in collision with inert body or system of reference from in front to the direction of the motion. The similar characterization can be given for the energy balance for collision from behind.

A.10.3.3. The proof of the energy transfer and additional frequency corrections S. A.10.3.3

With reference to 10H3, 10I7 and 10I8 the frequency of the collision from in *front* is:

$$f_{front}=f_a\left[1+\frac{v}{c}\right]=f_o\sqrt{\frac{1-(v/c)}{1+(v/c)}}\left[1+\frac{v}{c}\right]=f_o\sqrt{1-\frac{v^2}{c^2}} \quad \text{10L1}$$

For collision from *behind* is:

$$f_{behind}=f_a\left[1-\frac{v}{c}\right]=f_o\sqrt{\frac{1+(v/c)}{1-(v/c)}}\left[1-\frac{v}{c}\right]=f_o\sqrt{1-\frac{v^2}{c^2}} \quad \text{10L2}$$

from 10L1 and 10L2: $$\sqrt{\frac{1-(v/c)}{1+(v/c)}}\left[1+\frac{v}{c}\right]=\sqrt{1-\frac{v^2}{c^2}}=\sqrt{\frac{1+(v/c)}{1-(v/c)}}\left[1-\frac{v}{c}\right] \quad \text{10L3}$$

once $$\sqrt{\frac{1-(v/c)}{1+(v/c)}}=A=\sqrt{\frac{1+(v/c)}{1-(v/c)}\cdot\frac{1-(v/c)}{1+(v/c)}}; \quad \text{10L4}$$

$$\sqrt{\frac{1+(v/c)}{1-(v/c)}}=\frac{1}{A}=\sqrt{\frac{1-(v/c)}{1+(v/c)}\cdot\frac{1+(v/c)}{1-(v/c)}} \text{ indeed.} \quad \text{10L5}$$

10L4 and 10L5 prove that the substitution in 10K4 and 10K5 was correct. Ref. 10L1 10L2

10L1 and 10L2 well demonstrate the frequency correction function of δ_f in both cases:

- Either the system of reference in collision from in front gives off energy: as consequence of the decrease of the energy the speed of the motion is decreasing, which results in the increase of the frequency.
  With reference to 10L1, the $\delta_{f+}=1+(v/c)$ correction ensures the harmony with the time flow.
- Or the system of reference in collision from behind receives energy and speeds up: The increasing speed, with reference to 10L2 decreases the frequency and the coefficient of the correction $\delta_{f-}=1-(v/c)$ ensures the harmony with the time flow.

Here is the point when we have to return to the definition of δ_f, the coefficient of the correction. The correction is necessary in both directions: either it is about increasing the impact or decreasing it. And the format of the coefficient depends on, which side of the energy balance is to be corrected: Ref. 10I3-10I6

Either $\delta_{f+}=1+\frac{v}{c}$; and $\delta_{f-}=1-\frac{v}{c}$; or $\delta_{f+}=\frac{1}{1+(v/c)}$; or $\delta_{f-}=\frac{1}{1-(v/c)}$; 10L6

The wavelength correction $\lambda=1/f$ obviously needs reciprocal values.

This is the reason, there is a need to add an additional *fourth* comment to the concerns of *Einstein's* definitions in Section A.10.1.2:

4./
The coefficient Einstein suggests for the correction of the impact of the speed of light (?!) is necessary only to use in the case of all electromagnetic collusions for ensuring the harmony of the speed value, the time flow and the frequency.

S. A.10.4

A.10.4
Review of the principles of the foundation of the General Theory

The foundation of the General Theory is based on:

- an incomplete time formula for relative motion;
- a misunderstanding of the relation concerning the space coordinates;
- an *a priori* statement on gravitation, resulting in the unnecessary correction of the speed of light;

Einstein introduces the General Theory in Chapters 1-4 of 'C', still without addressing properly the reciprocal character of relativity.

In the case of having two systems of reference, 1 and 2 in relative motion with $v = const$, the so-called *Minkowski* space-time interval can be expressed in two ways, depending on our free choice on the supposed to be stationary systems of reference:

- either: - or:

10M1 $t_2^2c^2 = t_1^2c^2 - t_1^2v^2$ $t_1^2c^2 = t_2^2c^2 - t_2^2v^2$

In 10M1 t_1 and t_2 donate the time, measured in the systems of reference accordingly.

Because of this reciprocal character, the *de facto physical* distance between the systems, addressed in the equations of 10M1 (the systems in each case in relative motion make) is

10M2 the same: $D = t_1v$ and also $D = t_2v$.

10M3 Substituting 10M2 into 10M1 $t_2^2c^2 = t_1^2c^2 - D$; and $t_1^2c^2 = t_2^2c^2 - D$;

Meaning: Systems of reference in relative motion are equal in their relative status.
If there is no basic platform for comparison, the relativity may lead to justified paradox, where all components of the relative motion are reciprocal and ultimately equal.

S. A.10.4 1

A.10.4.1. Review of the principle of Einstein's space-time continuum

In a space which is free of gravitational fields, we are taking two systems of reference as *Einstein* advises in Chapter 3 of 'C': a Galilean system of reference K_o and also K, a system of reference in uniform rotation relative to K_o. The origin *'o'* of both systems and their axes of z_o and z permanently coincide.

Einstein insists that for space-time measurement in K, the system of reference in rotation, Euclidean geometry does not apply. We will show that, while space coordinates and time measurement cannot indeed be *projected* in conventional way, the main concern is not about the tools of projection rather the approach used.

The two systems of reference in the example are identical and congruent at rest. They are not taken as rigid discs or solid inert bodies, rather systems of reference with infinite number of individual inert mass points at loose at positions or coordinates within the systems of reference. The energy relation between mass points within the system of reference, the origin and the force of the rotation are not identified.

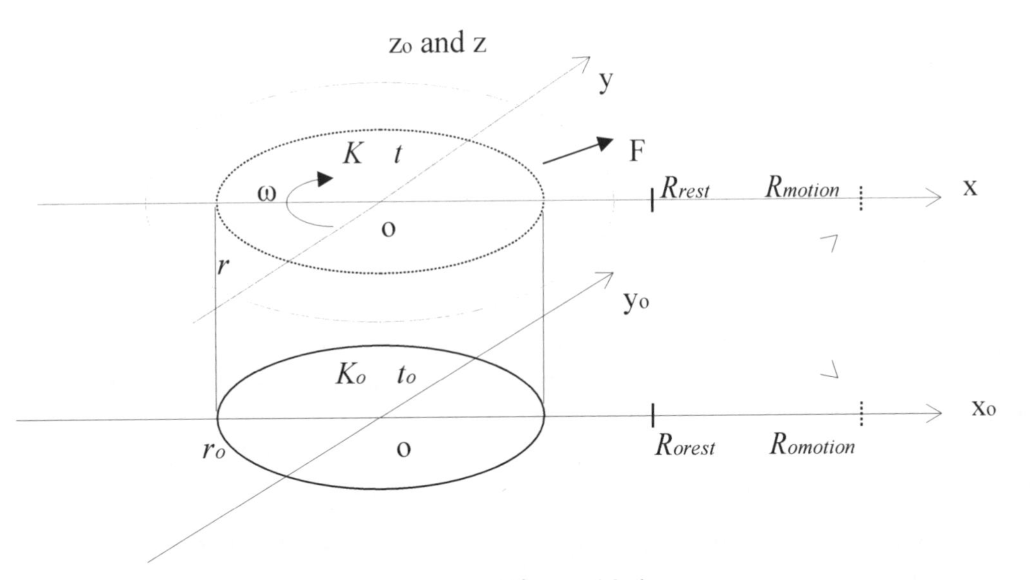

Figure 10.4 Fig. 10.4

The systems of reference are taken in Figure 10.4 as two-dimensional ones, with extension of the space in directions of axes x_o, x and y_o, y. The systems have no extension in direction of axes z_o and z.

The *first* concern about *Einstein's* example relates to the definition of the motion.

Independently of whether a stationary system like K_o does exist at all, or does not, the rotation is *not uniform*. The values of the angular speed of the rotation of K, measured within K_o and measured within K itself are different.
Why?
Because v, the relative speed between the coordinates or positions of the two systems of reference in relative motion must be reciprocal. But R, the measured distance of a radius of and in the system of reference in motion at peripheral speed v of the rotating K differs from that of R_o, of the same rotation of K, but measured within K_o, the stationary system of reference. The angular speed of the position or the coordinate with peripheral speed v, of

the rotation of K measured in K is $\omega_R = \frac{v}{R}$ 10N1

while the angular speed, measured in K_o is $\omega_o = \frac{v}{R_o}$ and $\omega_R \neq \omega_o$ 1N2

As usual, index 'o' denotes the parameter of the stationary system of reference.

The *second* concern is that the reciprocal character of the motion is missing.

Without energy considerations of any kind, there is no reason to consider *a priori* one of the systems as *stationary* and the other as in motion, while both are equal parts of the same relation.

When the two systems are at rest relative to each other, and R_o=R, the circumferences belonging to R_o and R are equal and the circles drawn to these radiuses are congruent.

There are fundamental differences between the two systems when they are in relative motion:

- Each and all coordinates of mass point positions of K_o, the stationary system of reference, are of the same time flow, since they all are equally at rest. Therefore, these are easily comparable with each other.
- *K*, the system of reference in rotation, is made up of an infinite number of systems of reference of mass point positions, which are of different motion and time flow. The comparison of space coordinates of these systems of reference *must be* based on motion.
  - positions on the circumferences, taken with zero radial extension, on equal radius of the rotation are parts of the same system of reference and time flow;
  - positions on the straight line, taken with zero extension, that connects the centre of the rotation with any coordinates of any circumferences of the rotation are in fact different systems of reference with different time flow;
  - the centre of *K*, the system of reference in rotation is at rest, and, therefore, its time system is equivalent to that of K_o, the stationary system of reference.

Ref. The distinguishing tool between the systems of reference above is the time relation:

10D11
10O1

$$dt = \frac{dt_o}{\sqrt{1 - \frac{v^2}{c^2}}}$$

where
t_o is time measurement at the centre of the system of reference in rotation and at all positions of the stationary system of reference;
v means the *peripheral velocity of the rotation* of *K* on a certain radius;
t is the time measurement in the system of reference at radius with peripheral speed of v, part of *K*.

Distances measured in the systems of reference in motion are a function of the time flow. Since all positions of a radius of the rotating motion are of different peripheral speed, all distances measured at circumferences of different radiuses are different.

We are taking v, a peripheral speed on a circumference. It identifies a certain R radius, measured from the centre of the rotation within one of the systems of reference of mass point positions in rotation, with peripheral speed of $v = \omega_R R$. (10O2)

We can also identify this as a circumference taken at radius R, measured within the system of reference in rotation at peripheral speed v.

We are taking the time period, during which the system of reference in rotation makes a complete spin, and denote this duration in the system of reference in rotation at the circumference of radius R as Δt_R.

10O3 From the transformation equation we receive $c^2 \Delta t_o^2 = c^2 \Delta t_R^2 - v^2 \Delta t_R^2$

10O4 Obviously $v = \omega_R R$ and $c^2 \Delta t_o^2 = c^2 \Delta t_R^2 - \omega_R^2 R^2 \Delta t_R^2$

10O5 The angular velocity at this circumference of radius R is: $\omega_R = \dfrac{2\Pi}{\Delta t_R}$

10O6 having 10O4 been substituted, getting: $R = \dfrac{c}{2\Pi} \sqrt{\Delta t_R^2 - \Delta t_o^2}$

(R is measured within *K*, the system of reference in rotation at position or coordinate of mass point with peripheral speed v; ω_R is the angular speed of the rotation at this radius, measured within the system of reference in rotation; Δt_o is the period of the time measured in K_o, the stationary system of reference, during the system of reference at a circumference with speed v of *K* makes a full spin.)

The meaning of the formula in 1O6 is:
This is the position of the system of reference of a mass point in motion, which completes a full cycle in rotation for time period Δt_R. It is measured in the system of reference in motion at *distance* R from the centre of rotation.

From 10O4 follows that $$\Delta t_o = \frac{1}{c}\sqrt{c^2\Delta t_R^2 - D^2\Pi^2} \qquad \text{10O7}$$

and since $\Delta t_R = \dfrac{2\Pi}{\omega_R}$; and $$\Delta t_o = \frac{1}{c}\sqrt{c^2\left(\frac{2\Pi}{\omega}\right)^2 - (2R)^2\Pi^2} = \frac{1}{c}\sqrt{\frac{4\Pi^2c^2}{\omega^2} - 4R^2\Pi^2}\ ; \qquad \text{10O8}$$

$$\Delta t_o = \frac{2\Pi}{c}\sqrt{\frac{c^2}{\omega_R} - R^2} \qquad \text{10P1}$$

Since the peripheral speed at radius R of the rotating system of reference is

$$v = \omega_R R; \qquad \Delta t_o = \frac{2\Pi}{c}R\sqrt{\frac{c^2}{v^2} - 1}\ ; \qquad \text{10P2}$$

With reference to 10D11, the time relation between the system of reference at the circumference of the rotating K at R distance, measured within K at peripheral speed of v, and the stationary system of reference is:	$\Delta t_R = \dfrac{\Delta t_o}{\sqrt{1 - \dfrac{v^2}{c^2}}}$

10P3

10P1 and 10P3 must be satisfied at the same time $$\frac{2\Pi}{c}R\sqrt{\frac{c^2}{v^2} - 1} = \Delta t_R\sqrt{1 - \frac{v^2}{c^2}}\ ; \qquad \text{10P4}$$

which gives the radius:

$$R = \frac{c\Delta t_R}{2\Pi}\sqrt{\frac{1 - \frac{v^2}{c^2}}{\frac{c^2}{v^2} - 1}} = \frac{c\Delta t_R}{2\Pi}\sqrt{\frac{\frac{c^2 - v^2}{c^2}}{\frac{c^2 - v^2}{v^2}}} = \frac{v\Delta t_R}{2\Pi}; \qquad R = \frac{v\Delta t_R}{2\Pi} = \frac{v\Delta t_o}{2\Pi}\cdot\frac{1}{\sqrt{1 - \frac{v^2}{c^2}}}; \qquad \text{10P5}$$

measured in the system of reference in motion.

Meaning of 10P5: <u>*The motion expands the space!*</u>

From 10P5 follows that if $v = 0$ $\quad R = 0$ and $t_o = \Delta t_R$

if $\lim v = c$ $\quad R = \infty$

The meaning of 10P1 fundamental:
10P1 gives the time relation of systems of reference in relative motion:

Substituting into 10P1: $$v = \omega_R R; \quad \omega_R = \frac{2\Pi}{\Delta t_R}; \quad R = \frac{v\Delta t_R}{2\Pi}; \quad \Delta t_o = \frac{2\Pi}{c}\frac{v\Delta t_R}{2\Pi}\sqrt{\frac{c^2 - v^2}{v^2}} \qquad \text{10P6}$$

Resulting: $$\Delta t_o = \Delta t_R\sqrt{1 - \frac{v^2}{c^2}} \qquad \text{10P7}$$

10P7 is in full compliance with 10D11: the motion speeds up the time flow.
It proves, contrary to Einstein's time formula that the time flows faster in systems of reference in motion! The strength of this deduction is that only the transformation equations and the tools of Euclidian geometry were used.
1P4 gives the meaning for the time: ***no motion (no event) = time cannot be defined***:

Substituting into 10P4 $$\omega_R = \frac{v}{R}; \quad \Delta t_o = \frac{2\Pi}{c}R\frac{\sqrt{c^2 - v^2}}{v}; \quad \Delta t_o = \frac{2\Pi}{c}\cdot\frac{1}{\omega_R}\sqrt{c^2 - (w_R R)^2}\ ; \qquad \text{10P8}$$

10P8 allows $R = 0$, but $\omega_R \neq 0$!

If $\omega_R = 0$, 10P8 has no meaning. Consequently, time parameters can only be defined if motion (event) is present.

The motion, which expands the space, may provide the reason and origin of the centrifugal force of rigid inert subjects of systems of reference in rotation.

For demonstration of the space expanding character of the motion, an Euclidian space position or coordinate of a mass point, measured at distance R_{orest} from the centre of K_o, the stationary system of reference, is taken in Figure 1.4. Once K has been in rotation, the respective R_{motion} space-coordinate within K, which corresponds to R_{orest} at status of rest, reference to 10O5, will have a distance to be measured from the centre of its rotation as

10P9 $$R_{motion} = \frac{c}{2\Pi}\sqrt{\Delta t_R^2 - \Delta t_o^2}\,; \quad \text{we can write 1X2 as} \quad R_{motion} = \frac{c\Delta t_o}{2\Pi}\sqrt{\frac{\Delta t_R}{\Delta t_o} - 1}$$

$$\text{and} \quad \Delta t_R > \Delta t_o \quad R_{motion} > R_{rest}$$

R_{motion} is the expanded radius measured in the system of reference in motion

10P10 $$\Delta t_R = t_R - 0 \quad \text{and} \quad \Delta t_o = t_o - 0$$

S. A.10.4 2

A.10.4.2. Dealing with the reciprocal character of the exercise above.

Both systems of reference are congruent at rest. The relation between them is the same, just now K_o is considered as being in rotation relative to the stationary K. All deductions above are the same, but the status of coordinates at R_{orest} and R_{motion} will be changed.

The reciprocal character of relative motion with speed v must be appreciated even if it is a consequence of a rotation, with the comment that *relative rotation* of two systems of reference as such is not uniform and hardly reciprocal.

Ref. Fig. 10.4

In the case of K in rotation, reference to Figure 10.4:

- coordinate R_{orest} measured within the stationary system of reference K_o has been in relative motion with speed v to coordinate R_{motion} measured in the rotating K;

Ref. Fig. 10.4

In the case of K_o in rotation, reference to Figure 10.4:

- coordinate R_{orest} within the stationary system of reference K has been in relative motion with speed v to coordinate of $R_{omotion}$ measured in the rotating K_o;

The relative speed between these coordinates of the two systems of reference is v in both cases, as it is expected to be in a reciprocal relation.

Relativity however is not all about reciprocal considerations, but also about effects of energy, work and force that make events happen in separate systems of reference in motion and at relative rest.

While the angular speed of the rotating system of reference, measured from the stationary system of reference is *uniform*, it is *non-uniform*, when measured from any position or coordinate of the rotating systems of reference or mass points.

With reference to 10P5, the expression for two different radiuses measured within two systems of reference of the variety of mass point positions in rotation

10R1 $$R_1 = \frac{v_1 \Delta t_{R1}}{2\Pi}\,; \quad \text{which gives} \quad \Delta t_{R1} = \frac{2\Pi}{\omega_{R1}}\,;$$

10R2 $$R_2 = \frac{v_2 \Delta t_{R2}}{2\Pi}\,; \quad \text{and} \quad \Delta t_{R2} = \frac{2\Pi}{\omega_{R2}}\,;$$

With reference to 10R1 and 10R2, the *non-uniformity* regarding the angular speed of the systems of reference in rotation are based on the tools (among them *П*) of the Euclidian geometry, which, because of the reciprocal character of the relation, hold good for both, stationary and systems of reference in rotating motion. (Planets are also orbiting around the *Sun*, measured on the *Earth*, with non-uniform angular speed.)

The motion expands the space and the gradient of the expansion can be written as $\frac{dR}{dt_o}$; 10R3

With reference to 10P5 and making the necessary substitutions:

$$R = \frac{vdt_R}{2\Pi}; \quad R = \frac{2\Pi}{\omega} \cdot \frac{v}{2\Pi} = \frac{v}{\omega}; \quad \text{where obviously } \Delta t_R = \frac{2\Pi}{\omega}; \qquad \text{10R4}$$

What is the character of the expansion of the space reference to the motion?

$$\text{since } \omega = const; \quad \frac{dR}{dt_o} = \frac{d\left(\frac{v}{\omega}\right)}{dt_o} = a \quad = \text{acceleration} \qquad \text{10R5}$$

As a conclusion can be stated, that the modified time formula; the modified understanding of space coordinates; the energy balance and the correction of frequencies and lengths of electromagnetic waves in collision; the motion that modifies the time flow and expands the space drive us towards the following recognition: acceleration is a valid alternative for changing of our existing views on gravitation.